D1530850

Advances in Physical Geochemistry

Volume 5

Editor-in-Chief

Surendra K. Saxena

Editorial Board

L. Barron P.M. Bell N.D. Chatterjee R. Kretz
D.H. Lindsley Y. Matsui A. Navrotsky R.C. Newton
L.L. Perchuk R. Powell R. Robie A.B. Thompson
B.J. Wood

Advances in Physical Geochemistry

Series Editor: Surendra K. Saxena

WITHDRAWN

552.4
F673w

Fluid–Rock Interactions during Metamorphism

Edited by
John V. Walther and Bernard J. Wood

With Contributions by
M.L. Crawford   J.M. Ferry   R.T. Gregory
L.S. Hollister   R.C. Newton   J. Ridley
A.B. Thompson   J.V. Walther   B.J. Wood
B.W.D. Yardley

With 59 Illustrations

Springer-Verlag
New York Berlin Heidelberg Tokyo

WITHDRAWN

John V. Walther
Department of Geological Sciences
Northwestern University
Evanston, Illinois 60201
U.S.A.

Bernard J. Wood
Department of Geological Sciences
Northwestern University
Evanston, Illinois 60201
U.S.A.

Series Editor
Surendra K. Saxena
Department of Geology
Brooklyn College
City University of New York
Brooklyn, New York 11210
U.S.A.

Library of Congress Cataloging in Publication Data
Main entry under title:
Fluid-rock interactions during metamorphism.
   (Advances in physical geochemistry; v. 5)
   Bibliography: p.
   Includes index.
   1. Metamorphism (Geology)   2. Water, Underground.
I. Walther, John V.   II. Wood, B.J.   III. Crawford, M.L.
IV. Series.
QE475.A2F597   1986   552'.4   85-27899

© 1986 by Springer-Verlag New York Inc.
All rights reserved. No part of this book may be translated or reproduced in any
form without written permission from Springer-Verlag, 175 Fifth Avenue, New
York, New York 10010, U.S.A.
The use of general descriptive names, trade names, trademarks, etc., in this
publication, even if the former are not especially identified, is not to be taken as a
sign that such names, as understood by the Trade Marks and Merchandise Marks
Act, may accordingly by used freely by anyone.

Typeset by Bi-Comp, Incorporated, York, Pennsylvania.
Printed and bound by R.R. Donnelley & Sons, Harrisonburg, Virginia.
Printed in the United States of America.

9 8 7 6 5 4 3 2 1

ISBN 0-387-96244-1 Springer-Verlag New York Berlin Heidelberg Tokyo
ISBN 3-540-96244-1 Springer-Verlag Berlin Heidelberg New York Tokyo

Preface

The fifth volume in this series is focused on the chemical and physical interactions between rocks undergoing metamorphism and the fluids that they generate and that pass through them. The recognition that such processes can profoundly affect the course of metamorphism has resulted in a number of recent papers and we consider that it is time for a review by some of the interested parties. We hope our selection of contributors provides an adequate cross section and demonstrates some of the flavor of this rapidly developing field.

A cursory examination of the volume will reveal that there are widely divergent opinions on the compositions of metamorphic fluids and on the ways in which they interact physically and chemically with the rocks through which they pass. Since our own views are extensively discussed in Chapters 4 and 8, we leave the reader to determine his own brand of the "truth."

We wish to thank D. Bird, S. Bohlen, D. Carmichael, G. Flowers, C. Foster, C. Graham, E. Perry, J. Selverstone, R. Tracy, J. Valley, and R. Wollast for their chapter reviews. Thanks are also due C. Cheverton for her editorial assistance, and the helpful staff at Springer-Verlag New York.

Evanston, Illinois

J.V. WALTHER
B.J. WOOD

ALLEGHENY COLLEGE LIBRARY

86-3358

ALLEGHENY COLLEGE LIBRARY

Contents

Contributors

CRAWFORD, M.L.   Department of Geology, Bryn Mawr College,
                     Bryn Mawr, Pennsylvania 19010 U.S.A.

FERRY, J.M.   Department of Earth and Planetary Sciences, The
                     Johns Hopkins University, Baltimore, Maryland
                     21218 U.S.A.

GREGORY, R.T.   Department of Earth Sciences, Monash
                     University, Clayton, Victoria 3168 Australia

HOLLISTER, L.S.   Department of Geological and Geophysical
                     Sciences, Princeton University, Princeton, New
                     Jersey 08544 U.S.A.

NEWTON, R.C.   Department of the Geophysical Sciences,
                     University of Chicago, Chicago, Illinois 60637
                     U.S.A.

RIDLEY, J.   Departement für Erdwissenschaften,
                     ETH-Zentrum, CH-8092, Zurich, Switzerland

THOMPSON, A.B.   Departement für Erdwissenschaften,
                     ETH-Zentrum, CH-8092, Zurich, Switzerland

WALTHER, J.V.   Department of Geological Sciences, Northwestern
                     University, Evanston, Illinois 60201 U.S.A.

WOOD, B.J.   Department of Geological Sciences, Northwestern
                     University, Evanston, Illinois 60201 U.S.A.

YARDLEY, B.W.D.   Department of Earth Sciences, University of
                     Leeds, Leeds LS2 9JT England

Chapter 1
Metamorphic Fluids: The Evidence from Fluid Inclusions

M.L. Crawford and L.S. Hollister

Introduction

Detailed microscopic studies of rocks from virtually all terrestrial environments show that one or more minerals in these rocks contain small quantities of liquid (or glass) and/or vapor trapped in cavities generally less than 50 microns in diameter. These are fluid inclusions, and they provide us with samples of the fluid phase present in the rock at some time during its evolution. Early petrographers (Sorby, 1858; Rosenbusch, 1923; Zirkel, 1873) included descriptions of these features. More recently, several individuals (notably Lemmlein, Dolgov, and Ermakov in the U.S.S.R.; Poty, Touret, Touray, and Weisbrod in France; and Roedder in the United States) have emphasized the petrologic information that can be obtained from these fluid inclusions. Two recent publications synthesize and summarize much of the previous work on fluid inclusions: the Mineralogical Association of Canada short course handbook *Fluid Inclusions: Applications to Petrology* (edited by Hollister and Crawford, 1981) and the Mineralogical Society of America Review in Mineralogy entitled *Fluid Inclusions* (Roedder, 1984). In this chapter, we emphasize important contributions toward understanding metamorphic processes that have been established through study of fluid inclusions, and we review several areas of ongoing research in which the study of fluid inclusions contributes to discussions of unresolved problems.

Application of fluid inclusion studies to metamorphic rocks in the United States has lagged somewhat because of concerns that have been expressed about the relevance of those studies to the identification of fluids present during metamorphism. It was recognized early (Tuttle, 1949) that most fluid inclusions occur on fractures that cross grain boundaries and therefore could have been trapped well after the metamorphic event, perhaps at near-surface conditions. The discovery by Touret (1971) of dense liquid CO_2 inclusions (i.e., not typical groundwater) opened the door for much of the recent research. Roedder (1981a, 1981b, 1984) has emphasized the small size of fluid inclusions in metamorphic rocks and the possibility that exsolution or selective leakage of fluid components through the host quartz could modify the compositions and densities of those small inclusions, especially considering the long time periods that metamorphic rocks are close to their maximum $P-T$ conditions. In this chapter, we review information obtained from fluid inclusion studies in metamorphic rocks, and we conclude that fluid inclusions do provide information on one or more intervals in the history of a metamorphic rock between peak temperatures and the surface; the cumulative body of empirical data suggests that Roedder's concerns do not apply for most reported metamorphic fluid inclusion studies.

Trapping Mechanisms of Fluid Inclusions in Metamorphic Rocks

Some fluid inclusions are incorporated during crystal growth at irregularities on the surface of the growing mineral. These fluid inclusions, termed *primary* by Roedder (1972), occur commonly in mineral grains that grow in an environment with abundant fluid, for example in hydrothermal veins, in igneous rocks, or in some diagenetic cements. Other fluid inclusions form by sealing and healing of fluid-filled fractures in minerals (Shelton and Orville, 1980; Smith and Evans, 1984). The walls of these fractures grow together, isolating small pockets of the fluid that had entered the open crack. These are called *secondary inclusions,* and they are common in rocks with low porosity or in environments in which grains are subject to tectonic or to thermal stresses during or after growth. Secondary inclusions along healed cracks form after the host mineral grain crystallized, although in some cases grain growth could have been proceeding when the crack opened.

Both types of inclusions, those trapped during crystal growth and those trapped along healed microcracks, occur in metamorphic rocks. In metamorphic veins, especially veins formed in rocks not subject to subsequent deformation or recrystallization, primary inclusions may be found in the metamorphic minerals (e.g., Stalder and Touray, 1970; Poty *et al.*, 1974; Mullis, 1976). Primary inclusions have also been reported from garnets in skarn (Tan and Kwak, 1979). These probably result from rapid growth, in a fluid-rich environment, of a mineral that commonly contains numerous inclusions. In most metamorphic rocks, however, as well as in syntectonic veins and pods, crack formation and healing is the most common and, in many cases, the only mechanism for trapping fluid inclusions.

Walther and Orville (1982) have described possible mechanisms of volatile transport in metamorphic rocks undergoing devolatilization. They calculate that the fluid flux that could be accommodated by flow on grain boundaries is several orders of magnitude lower than the estimated flux required by devolatilization of an average pelite. On the other hand, fluid-filled fractures could accommodate the flux. These fractures propagate upward through a rock driven by the density difference between fluid and host rock at conditions where volatiles are being generated by metamorphic reactions and the rock tensile strength is low. By their calculations, lengths of such fractures range from 10^2 to 10^{-2} cm for each cm^2 area of rock, depending on their width. Presumably, the numerous healed fractures marked by trails of fluid inclusions record the traces of these fractures.

Observations, experiments, and calculations thus suggest that fluids in metamorphic rocks are transported along fractures ranging in size from microcracks to veins and that planes of fluid inclusions (the so-called secondary inclusions) mark the healed traces of these conduits. It is very likely that fluid in many of these inclusions originated by devolatilization reactions in the immediately surrounding rock and was in equilibrium with the host rock mineral assemblage at the time of formation of the inclusions. Such inclusions preserve the same information as primary inclusions formed during mineral growth. These mechanisms for fluid movement also permit fluids percolating through metamorphic rocks to be trapped at some distance from the fluid source or significantly later than the host rock metamorphic mineral assemblage (Yardley, 1983). This latter situation is most likely for fluid inclusions trapped in minerals in cross-cutting veins but cannot be ruled out for rocks subject to pervasive influx of fluid from external sources. To avoid the ambiguity of applying definitions adapted for hydrothermal veins to fluid inclusion occurrences in metamorphic rocks, we find it useful to designate successive generations of secondary fluid inclusions trapped along fractures according to their time sequence of development. In our experience, the earliest generation is most likely to have equilibrated with the host rock at maximum temperature conditions, and the later generations represent sequential stages in the retrograde thermal history of the rock.

Postentrapment Modification of Fluid Inclusions

During prograde metamorphism, fluid inclusions trapped at low temperatures are generally destroyed as the temperature of the rock rises. This results from recrystallization of the host minerals caused by metamorphic reactions and by thermally driven grain coarsening and elimination of lattice defects. Even if early fluid inclusions persist, fluid densities will be modified by thermal expansion of the fluid in all settings in which the gradient of temperature increase is flatter than the slope of the fluid isochores (curves of constant molar volume or density) (Fig. 1). In most metamorphic environments, rocks respond to pressure changes significantly more rapidly than to temperature changes. Thus, it can be expected that a rock will reach its metamorphic maximum pressure before the maximum temperature is achieved, yielding a prograde metamorphic $P-T$ path concave toward the temperature axis. This type of path crosses aqueous fluid isochores at a high

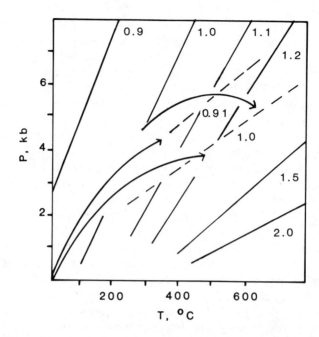

Fig. 1. Representative isochores for H_2O (solid) and CO_2 (dashed), labeled with molar volumes. The arrows show several schematic prograde metamorphic paths chosen to illustrate the fact that the P–T path for metamorphic rocks crosses both types of isochores, from lower to higher specific volume, at a high angle near peak metamorphic conditions.

angle close to the thermal peak of metamorphism (Fig. 1; see also Norris and Henley, 1976), resulting in significant increase in the molar volumes of fluid inclusions trapped below peak metamorphic temperatures. Because fluid thermal expansion is greater than that of the solid surrounding the inclusion, a pressure in excess of confining pressure will develop within the inclusion cavity.

The maximum excess pressure that can occur within an inclusion, defined as pressure within the inclusion minus confining pressure on the host mineral, is limited by the strength of the host mineral. For inclusions in quartz, laboratory experiments by Naumov *et al.* (1966), confirmed by Leroy (1979), determined that the maximum pressure that quartz can contain at room temperature increases as the inclusion size decreases below about 12 μm in diameter. Inclusions above 12 μm rupture when the excess pressure exceeds 850 bar. Those below 12 μm can contain pressures of 1.5 kb or more, depending on shape of inclusion and, presumably, imperfections in the host quartz near the inclusion. Swanenberg (1980) reported an excess pressure of 6 kb for a 1-μm inclusion.

The instantaneous laboratory decrepitation observations probably give maximum pressure differences that can be sustained by quartz; excess pressures at high temperature for a lengthy period of time must increase the ease with which quartz deforms around the inclusion. Burruss and Hollister (1979) concluded, based on study of a natural occurrence, that an excess pressure of about 1 kb is necessary to fracture 10-μm inclusions in quartz even if they are held at over 200°C for 10^5 to 10^6 yr. The fluid inclusions of this study are contained in samples collected from a 3-km core in crystalline rock near Los Alamos, New Mexico, where Precambrian rocks had been heated by the event that formed the Pleistocene Valles Caldera. Fluid inclusions were studied from samples collected at intervals down the core. Burruss and Hollister (1979) observed that below 1.5 km (Fig. 2), pressure-corrected temperatures of entrapment corresponded approximately to measured downhole temperatures and that the inferred trapping temperatures increased regularly with depth. Homogenization temperatures of inclusions in samples from less than 1.5 km were highly variable and showed no systematic relation to depth. These shallow samples contained many one-phase water inclusions. From Fig. 2 it can be seen that at a depth of 1.5 km, one-phase pure H_2O inclusions with a density appropriate for trapping at shallow depths would develop an excess pressure of 1 kb when heated to the temperature observed at that depth. The conclusion advanced by Burruss and Hollister (1979) was that these dense inclusions initially trapped at low temperatures became overpressured and broke or leaked. As they did so, they reopened fractures containing other inclusions, homogenizing the composition of fluid within the fracture and resetting the density. A new set of secondary inclusions formed when the fracture rehealed. Very small inclusions in the deep samples were unsuitable for study, making it impossible to confirm if excess pressures greater than 1 kb could be preserved in this environment in the small inclusions.

At high temperature and pressure, deformation, accomplished by either plastic flow or by moving material away from the inclusion walls to the grain boundary by dislocation climb, would result in volume changes and hence changes of density of fluid inclusions without necessarily rupturing or leaking. The laboratory data which show that small inclusions can sustain higher internal pressures than large inclusions without leaking do, however, seem to apply also to natural settings, particularly at low and medium grades of metamorphism. In addition, studies of several generations of fluid inclusions in single metamorphic rocks report internal pressure differences of 1–2 kb between early inclusions and later ones at temperatures as high as 300–400°C (e.g., Hollister et al., 1979; Sisson et al., 1981; Crawford et al., 1984; Selverstone et al., 1984). Models for fluid flow such as that presented by Walther and Orville (1982) should take account of this evidence from fluid inclusions which suggests that quartz and other minerals that preserve fluid inclusions must have some finite strength to retain these small volumes of "overpressured" fluid.

Several generations of fluid inclusions may result from episodes of fracturing and leakage of inclusions in response to pressure differences between the inclusion fluid and its surroundings, as discussed above. In the labora-

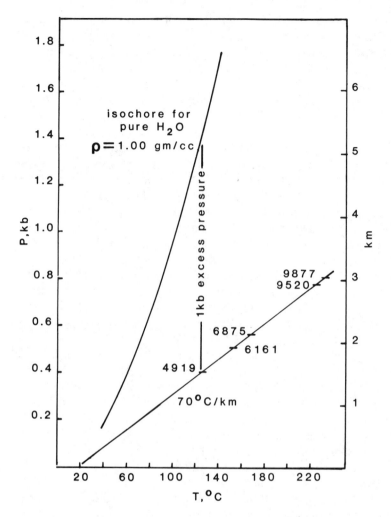

Fig. 2. Results of the fluid inclusion study at Valles Caldera, New Mexico (Burruss and Hollister, 1979). The lower curve connects the homogenization temperatures for samples (numbered) collected at the depths indicated on the right-hand ordinate. Pressures inside pure water inclusions trapped at near-surface conditions are constrained by the isochore for H_2O with density of 1.0. At depths greater than 1.5 km an excess pressure >1 kb caused the pure water inclusions to rupture.

tory, inclusions decrepitate by formation of microcracks that connect the inclusion to the sample surface. The fluid leaks out along the fracture. Under confining pressure, in a geologic setting, excess internal pressure probably opens a microcrack that may, but need not, reach the grain boundary. The crack volume increases until confining pressure and pressure in the crack are equal. Thereafter, the crack heals, forming a planar array of new fluid inclusions. Sudden rupturing of early inclusions, accompanied by healing of the grain around the exploded inclusion, results in three-dimensional arrays called *decrepitation clusters,* which resemble starbursts. In this case, the fluid is retrapped along tiny fractures generated during the decrepitation. These have been noted by Lemmlein (1956), Bilal and Touret (1976), Swanenberg (1980), Kreulen (1980), and Hollister (1982), among others.

Fluid Inclusions as Indicators of Metamorphic Fluids

Although fluid inclusions have been reported in many metamorphic minerals including garnet, pyroxene, andalusite, kyanite, corundum, epidote, feldspars, scapolite, tremolite, cordierite, carbonates, apatite, and tourmaline, they are most commonly found only in quartz. In many ways, quartz is an ideal mineral host for the study of metamorphic fluid inclusions, because it occurs in most metamorphic rocks and is stable at all metamorphic conditions within the crust. In addition to forming an integral part of the matrix assemblage of metamorphic rocks, quartz is readily mobilized by fluids and reprecipitated in veins and pods. Thus, studying inclusions in quartz permits us to examine the fluids present in the metamorphic rocks as well as to investigate the fluids percolating along cracks and fissures, generating the quartz segregations. Quartz does not react chemically with the trapped fluid, in contrast to other minerals that could exchange cations with species dissolved in aqueous fluid inclusions. Thus, we can assume that compositions of fluid inclusions trapped in quartz are preserved provided the inclusions do not leak.

Identification of Metamorphic Fluids: Primary Inclusions

An aim of fluid inclusion studies in metamorphic rocks is correlation of particular fluids with specific $P-T$ conditions. One way to identify metamorphic fluids involves finding fluid inclusions trapped during grain growth (primary inclusions) in metamorphic minerals diagnostic of those $P-T$ conditions. The densities of fluids in primary fluid inclusions in these minerals correspond closely to those predicted from the metamorphic conditions (Berglund and Touret, 1976; Tan and Kwak, 1979; Coolen, 1980).

By itself, quartz is one of the poorest indicators of metamorphic conditions. Since quartz participates in most metamorphic reactions, many of which are quartz consuming, it is impossible to link growth of a quartz grain in a schist or gneiss with a specific metamorphic event. Quartz also recrystallizes readily, especially after it has been strained, leading to possible changes in the volume and in the redistribution of preexisting inclusions (Kerrich, 1976; Wilkins and Barkas, 1978). This process may obliterate early formed fluid inclusions; however, it allows use of inclusions to deduce conditions during annealing of the quartz, which may be as important a piece of information in studying the history of a metamorphic rock as identifying the fluid present at peak metamorphic temperatures. Where quartz grain growth can be closely related to growth of other minerals, for example in hydrothermal veins, conditions of quartz growth may be inferred from associated phases (e.g., Saliot *et al.*, 1982). In this situation, primary fluid inclusions in quartz may be linked to the temperature and pressure regime deduced from the associated minerals.

Limits may also be set on timing of fluid inclusion formation if they are trapped in quartz grains subsequently incorporated as solid inclusions in a poikilitic mineral such as garnet. Hollister and Burruss (1976) observed that some of the inclusions in a group of high-grade gneisses that most closely fit the criteria for containing metamorphic fluid occurred in quartz inclusions in garnet. These inclusions presumably record conditions during garnet growth and were protected from subsequent disruption by the garnet.

Because unambiguous primary inclusions are so rarely identified in metamorphic rocks, other criteria must usually be applied in order to correlate a particular fluid inclusion population to a recognizable stage in the history of a metamorphic rock.

Identification of Metamorphic Fluids: Evidence from Fluid Densities and Textures

Information on the pressure and temperature conditions at which an inclusion or group of inclusions was formed may be inferred from the density of the fluid trapped in the inclusion. The $P-T$ point at which a fluid inclusion becomes a closed constant-volume system determines the density and thus the isochore for that fluid inclusion. Thus, determination of fluid density in an inclusion, derived from the temperature at which a fluid inclusion homogenizes to one fluid phase (Roedder, 1972; Touret, 1977) or calculated by summing densities of known volumes of fluid (Burruss, 1981), limits the possible trapping conditions to points along the relevant isochore. If a pressure and temperature bracket for metamorphism can be identified from the mineral assemblage and a group of fluid inclusions occurs with densities appropriate to those conditions, the data permit the interpretation that the fluid may indeed have been trapped at those metamorphic conditions.

Petrographic observations may allow determination of a sequence of generations of microcracks containing fluid inclusions (Roedder, 1984, chapter 12). In some cases, several generations of fractures can be identified by cross-cutting relations (Pagel, 1975; Touret, 1981), thus establishing a trapping sequence for the fluid inclusions along those healed fractures. Hollister and Burruss (1976) identified other petrographic features of a group of fluid inclusions that may suggest a trapping sequence. They noted several types of fluid inclusion occurrences within single samples: small, isolated inclusions; nonplanar groups or clusters; and inclusions along distinct planes.

Using density differences between these three types and the assumption that all inclusions preserved in metamorphic rocks were trapped at or after peak metamorphic temperatures as discussed above, they suggested that the isolated inclusions are the oldest, followed by the clusters, and finally by the inclusions along planes. Transposition and redistribution of trails of fluid inclusions away from the plane of original healed fractures has been noted by Roedder (1971) and Swanenberg (1980). These authors suggested that annealing and recrystallization of quartz around inclusions, accompanied by migration of inclusions along lines of weakness in the host quartz, may gradually obliterate the traces of older inclusion trails in a particular sample. Thus, the earliest formed inclusions are most likely those that do not occur along recognizable planes.

The texturally earliest inclusions identified using these petrographic criteria contain fluids with densities closest to those predicted from the metamorphic conditions. Hollister *et al.* (1979) showed that the earliest fluid inclusions from different metamorphic terranes differ in density as a function of metamorphic conditions; most have densities consistent with independently determined metamorphic temperatures and pressures. Other studies (e.g., Rich, 1979; Kreulen, 1980; Sisson *et al.*, 1981; Hollister, 1982; Yardley *et al.*, 1983; Crawford *et al.*, 1984; Ahrens, 1984) have confirmed this relation. Commonly, the inclusions identified texturally as being later also have lower densities than earlier ones (Weisbrod and Poty, 1975; Hollister and Burruss, 1976; Touret and Dietvorst, 1983). Important exceptions to this relation occur in situations of isobaric cooling (Tan and Kwak, 1979; Selverstone, 1982; Swanenberg, 1980; V.B. Sisson, 1985, personal communication). In this case, inclusions trapped at lower temperatures will be denser than those trapped at higher temperatures.

The combined use of textural and density criteria has permitted most students of fluid inclusions in metamorphic rocks to identify inclusions that formed during or close to peak metamorphic conditions and that have survived transport back to the surface. Samples that contain clearly primary fluid inclusions in minerals, such as garnet, that provide an indication of metamorphic conditions (Berglund and Touret, 1976; Tan and Kwak, 1979; Coolen, 1980) can be used to confirm inferences about identification of particular fluids with points along the *P–T* trajectory of a metamorphic rock.

Fluid Compositions

COH Fluids

The correspondence in composition of fluid inclusions, inferred from density determinations to be metamorphic fluids, with compositions predicted from petrologic and geochemical arguments, supports the conclusion that those fluids equilibrated with the minerals in the host rock and therefore represent samples of the metamorphic fluid. A recent study (Crawford *et al.*, 1984; Stout *et al.*, in press), specifically designed to evaluate this question, came to the same conclusion. The samples studied were selected from a few outcrops that were very well characterized petrographically. Table 1 taken from this study shows that the compositions of fluid inclusions with densities appropriate to the inferred metamorphic grade correspond closely to the compositions of fluids calculated from mineral compositions in the schists.

Inferences about metamorphic fluid compositions based on observations of fluid inclusions rely on the assumption that the $P-V-T-X$ relations of the fluids at the laboratory conditions at which measurements are carried out are close to the $P-V-T-X$ relations in those same fluids when they were trapped. This assumption has recently been addressed by Dubessy (1984) for COH fluids. If an inclusion remains a constant volume system after formation, he calculates that the volumetric properties and compositions of fluids in inclusions that do not contain a separate solid C phase (e.g., graphite) remain within a few percent of their equilibrium values at the trapping conditions. If graphite also occurs in the inclusion, either as a daughter mineral precipitated from the fluid or initially trapped simultaneously with the fluid, the same conclusion holds unless the initial fluid consisted of approximately equal numbers of C, O, and H atoms. In this latter case, chemical reequilibration may considerably modify (on the order of tens of percent) both the molecular species present in the inclusion and the density of the fluid. Because of this effect Dubessy suggests, for those inclusions that contain a separate C phase, that fluid compositions and densities at the conditions at which the fluid inclusions formed cannot be assumed to correspond to the values observed in the laboratory; they need to be calculated at the inferred trapping conditions.

Summaries of studies of fluid inclusions in metamorphic rocks from a variety of different grades and rock types show that there is a consistent

Table 1. Comparison of fluid compositions calculated from inclusion measurements and from mineral phase equilibria

Sample number		Observed			Calculated	
		X_{H_2O}	X_{CO_2}	X_{NaCl}	$X_{H_2O}{}^a$	$X_{H_2O}{}^b$
EG 2.3k	e	0.79	0.1996	0.0149		
	h	0.81	0.18	0.014		
	i	0.95	0.05	0.009		
					0.84	0.85
EG 2.3t	a	0.82	0.17	0.008		
	4	0.86	0.13	0.013		
	5	0.88	0.11	0.010		
	b	0.96	0.04	0.006		

[a] Fugacity coefficients from Jacobs and Kerrick (1981); f_{H_2O} from Chatterjee and Froese (1975).
[b] Fugacity coefficients from Jacobs and Kerrick (1981); f_{H_2O} from Ghent *et al.* (1982), Eugster *et al.* (1972).

relation of compositions of fluids found in inclusions to metamorphic conditions. Crawford (1981a), Touret (1981), and Roedder (1984) summarize variations in composition of fluid inclusions with metamorphic grade. These variations in fluid composition qualitatively agree with thermodynamic predictions. Although preferential diffusion of H_2O relative to CO_2 through or into quartz has been proposed (Roedder, 1984, p. 341) as a possible problem in metamorphic rocks, it has not been demonstrated and seems unlikely in view of the reported occurrences of H_2O + CO_2 inclusions with densities that correspond to high-grade metamorphic conditions (Hollister et al., 1979; Kreulen, 1980; Hollister, 1982; Sauniac and Touret, 1983; Crawford et al., 1984).

An impressive regional study across metamorphic isograds comes from work done on quartz veins in the Swiss Alps (Frey et al., 1980; Mullis et al., 1983). The metamorphic rocks that host the quartz veins range in grade from unmetamorphosed sediments to the amphibolite facies (mesozone). The fluids show a progression of compositions characterized by ~1 to >80 mol% higher hydrocarbons (HHC) in the unmetamorphosed rocks, ~1 to >90 mol% CH_4, <1 mol% HHC in the lower grade anchizone, ~70 to >99 mol% H_2O, <1 mol% CH_4 in the high-grade anchizone and epizone, and ~10 to >60 mol% CO_2 in the mesozone. In the Alps, the lower grade anchizone is defined by measurements of illite crystallinity and coal rank. The transition to the high-grade anchizone occurs approximately at the pyrophyllite isograd, and the mesozone starts at the staurolite isograd. On the Greek island of Naxos (Kreulen, 1980) and in the Himalayas (Pecher, 1979), fluid inclusions in greenschist and amphibolite facies rocks consist of CO_2 and H_2O, which is in agreement with the observations on rocks of similar metamorphic grade in the Alps as well as those in other areas. In the granulite facies, the fluids become dominantly CO_2 (Touret, 1974, 1981; Hansen et al., 1984; Newton, this volume).

The source of the methane in low-grade rocks is probably in large part from hydrocarbons more complex than methane. At higher grades, equilibria between dispersed carbon, iron-bearing silicate and iron oxide oxygen buffers, and H_2O and CO_2 produced by devolatilization control the composition of C-O-H fluids (Ohmoto and Kerrick, 1977; Frost, 1979; Holloway, 1981, 1984).

Because the compositions of C-O-H fluids are sensitive to f_{O_2} as well as temperature, fluid inclusion observations can be used to infer the oxygen fugacity for a metamorphic assemblage. French (1966), Eugster and Skippen (1967), and Frost (1979) point out that f_{O_2} values in graphitic rocks may be controlled by equilibria between H_2O released by devolatilization reactions and C. Such fluids must contain both CO_2 and CH_4 in approximately equal amounts; the total amount of carbon in the fluid phase increases with temperature. In contrast, the carbon species in fluids with f_{O_2} controlled by another buffer such as provided by iron-bearing silicate or iron oxides will be CH_4 at low temperatures and CO_2 at high temperatures. Thus, the H_2O-CO_2-CH_4 fluid reported by Hollister and Burruss (1976) from inclusions in graphitic granulite facies rocks suggests that H_2O-graphite equilibria played the major role in determining fluid composition and f_{O_2}, whereas the dominantly CO_2-rich fluid inclusions observed by Kreulen (1980) in graphitic sandstones from Naxos imply that graphite did not act as a buffer in that setting.

The numerous observed cases of correspondence in composition of texturally earliest formed fluid inclusions to compositions predicted for fluids buffered by the host mineral assemblages suggest that samples of primary fluids are preserved as fluid inclusions. However, complexities resulting from natural decrepitation (as discussed above), multiple stages of fluid entrapment, and fluid immiscibility (discussed below), make it difficult to define uniquely the primary fluid composition by microthermometry alone without corroborating evidence from the mineral assemblage.

Aqueous Fluids

Whereas the nature and relative amounts of species in the C-O-H system in fluid inclusions appear to depend strongly on metamorphic grade, the compositions of aqueous fluid inclusions depend in large measure on the compositions of host rocks and on mineral–fluid equilibria. Predicting the composition of aqueous fluids from mineral equilibria and thermodynamic data is not as straightforward as predicting the nature and abundance of species in COH fluids. This is due to the presence in the fluid phase of species, such as Cl^- and Br^-, that do not occur in major quantities in the associated mineral phases. Phase equilibria calculations fail in this circumstance.

Fluid inclusion studies have demonstrated that most aqueous fluids contain dissolved salts, based on melting temperatures of the fluid determined during freezing and heating studies. For brines, these compositions are reported as the wt% NaCl required to produce the observed brine melting temperature (percent NaCl equivalent). This can be justified because NaCl is the most abundant dissolved component in fluid inclusion brines, and other components do not affect the correlation between melting point and salinity to any great degree (Crawford, 1981b). Metamorphic fluids have salinities that range from values on the order of 2–6 wt% NaCl equivalent in pelitic schists and gneisses (Poty et al., 1974; Hollister and Burruss, 1976; Luckscheiter and Morteani, 1980) to 20–25 wt% NaCl equivalent in calcareous rocks (Crawford et al., 1979; Kreulen, 1980; Sisson et al., 1981), and even higher salinities where evaporites are present or suspected (Touray, 1970; Crampon, 1973; Saliot et al., 1978; Rich, 1979). The salinity of these fluids is controlled by the amount of dissolved anions, particularly chloride (Roedder, 1972). Observations of freezing point depression in aqueous fluid inclusions give good estimates of the salinities (Crawford, 1981b) that can then be used to calculate compositions of electrolyte solutions. However, the techniques of analysis for fluid inclusions in metamorphic rocks are not readily adapted to measuring accurately the nature and abundance of the dissolved salts. This is a result of the fact that most metamorphic rocks contain a variety of fluid inclusions of several compositions and hence bulk fluid extraction techniques cannot conveniently be used. Most spectroscopic techniques do not detect dissociated ions, and composition estimates based on freezing and melting data, the most routinely applied technique for determining compositions of fluids in inclusions, are not capable of giving more than a qualitative estimate of the amounts of major dissolved components (Crawford, 1981b).

A number of studies on fluid inclusions in metamorphic rocks (cited in Crawford, 1981a) show that the cation composition is controlled by the associated mineral assemblages. Thus, where the host rock contains primarily sodic and potassic phases, as in pelitic schists and gneisses and feldspathic rocks, the aqueous fluids contain Na and K, whereas in those cases in which calcic phases are more abundant, Ca is a major component of the dissolved salts. Work on equilibria between chloride brines and sedimentary and metamorphic mineral assemblages (Althaus and Johannes, 1969; Schulien, 1980; Bischoff et al., 1981; Eugster and Gunter, 1981; Vidale, 1983) predict that Na should be the dominant cation and Mg the least abundant. The relative amounts of K, Ca, and Fe depend on mineral assemblage, salinity, and temperature. SiO_2 is also present in solution in chloride brines, but we assume the dissolved SiO_2 is precipitated onto the walls of the fluid inclusions as temperature drops to surface conditions.

The compositions of aqueous fluids trapped in quartz probably do not change after trapping, as mentioned above. However, this may not be true for other mineral hosts. Observations we have made on inclusions trapped simultaneously in calcite and quartz show abundant dissolved Ca in the inclusions in calcite, but insignificant amounts of Ca in the aqueous brine inclusions in quartz. This suggests that estimates of dissolved salt compositions should be made cautiously if the mineral host to the fluid inclusion contains ions that will readily exchange in a chloride brine.

The documentation, from fluid inclusion studies, of high salinities in metamorphic fluids has several consequences. First, the studies record the presence of fluids that may be available for dissolving and transporting significant quantities of metals within metamorphic terranes. Chlorine-rich brines in diagenetic and hydrothermal settings have been identified as potential ore-forming fluids. The significant role of these brines in metamorphic processes suggests that further work is needed to provide data on brines at metamorphic conditions. Second, salt components play a significant role in expanding the limits of miscibility of mixed H_2O-CH_4 and H_2O-CO_2 fluids, as discussed below.

Additional Components

A further contribution to knowledge of the composition of the fluid phase has been the identification of nitrogen as a component, in some cases a major component, in the fluid phase of a wide range of metamorphic rocks (Swanenberg, 1980; Guilhaumou et al., 1981; Kreulen and Schuiling, 1982; Touret and Dietvorst, 1983; Yardley et al., 1983). These studies report nitrogen-bearing fluid inclusions from low-grade or unmetamorphosed dolomite, from high-grade gneisses, and from migmatites. We have also found nitrogen in greenschist and amphibolite grade samples (Stout *et al.*, in press). Although nitrogen is not nearly as abundant as H_2O, CO_2, and CH_4 in fluid inclusions, its presence shows that it cannot be ignored as a possible diluent for aqueous and carbonic fluids. The occurrence of nitrogen also poses a petrologic problem, namely a source for that component. In very low-grade schists, it is reasonable to suggest that nitrogen is released from organic compounds as they decompose thermally (Paxton, 1984). During this release, ammonium ions (NH_4^+) may replace K^+ or other exchangeable cations in clay minerals. Studies of higher grade igneous and metamorphic rocks (Wlotzka, 1961; Milovskiy and Volynets, 1966; Itihara and Honma, 1979) suggest several metamorphic minerals contain NH_4^+ replacing K^+. According to Wlotzka (1961), who analyzed coexisting minerals, biotite has the highest content of NH_4^+, followed in decreasing order by muscovite and potassium feldspar. The presence of nitrogen in fluid inclusions may thus reflect metamorphic processes resulting in the decomposition of those minerals, including breakdown during anatexis.

Fluid Immiscibility

Several metamorphic fluids are immiscible over a range of metamorphic conditions. These are mixtures of water or aqueous brine with C, S, or N-bearing compounds. H_2O-CO_2, H_2O-CH_4, and H_2O-N_2 mixtures all exhibit immiscibility at temperatures below 400°C (Franck, 1977) (Fig. 3). Until recently, neither experimental nor metamorphic petrologists were fully aware of the significance and consequences of this immiscibility because the consolute point in the pure H_2O-CO_2 system lies close to 270°C (Todheide and Franck, 1963), which is below most temperatures of metamorphic interest, and because little attention was paid to H_2O-CH_4–mineral equilibria. However, the presence of soluble salts, most commonly Cl salts in natural brines, increases the range of immiscibility both by raising the temperature of the consolute point and increasing the compositional range in which fluids unmix (Takenouchi and Kennedy, 1965; Naumov *et al.*, 1974; Gehrig, 1980).

Petrologic Role of Immiscible Fluids

Hollister (1973) presented one of the first discussions of the role of immiscible fluids in metamorphic rocks. Numerous subsequent studies, including Hollister and Burruss (1976), Mercolli (1979), Crawford *et al.* (1979), Pecher (1979), Kreulen (1980), Luckscheiter and Morteani (1980), Ypma and Fuzikawa (1980), Hendel and Hollister (1981), and Sisson *et al.* (1981), have

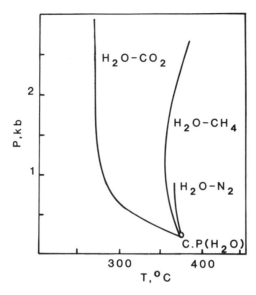

Fig. 3. Critical curves for the H_2O-CO_2, H_2O-CH_4, and H_2O-N_2 systems from Franck (1977). (Reprinted with permission from "Phase Equilibria and Fluid Properties in the Chemical Industry, ACS Symposium Series 60." Copyright 1977 American Chemical Society.) For each curve the two-phase region of immiscible fluids lies to the low temperature side.

documented or discussed fluid immiscibility during metamorphism. Bowers and Helgeson (1983a, 1983b) discuss in detail the effect of immiscibility in the H_2O-CO_2-$NaCl$ system on mineral equilibria. The presence of NaCl not only enlarges the H_2O-CO_2 solvus and raises the consolute point to temperatures above 500°C (Sisson *et al.*, 1981; Bowers and Helgeson, 1983a) but also increases the nonideality of H_2O-CO_2 fluid mixing. Mineral assemblages that are stable at low temperatures in the presence of water-rich fluid in the pure H_2O-CO_2 system occur at higher temperatures in the presence of brine. In addition, mineral assemblages may coexist with fluids over a much wider range of compositions in the H_2O-CO_2-$NaCl$ system than in the H_2O-CO_2 system. Equivalent isobarically invariant and univariant assemblages lie on both the H_2O-rich and CO_2-rich side of the solvus and are connected across the solvus by tie lines (Bowers and Helgeson, 1983b). As a consequence, mineral assemblages cannot be used to infer fluid compositions at conditions of immiscibility in the same way as in the pure H_2O-CO_2 system.

Studies of prograde metamorphism in calcareous rocks (e.g., Sanford, 1980; Granath *et al.*, 1983; Bowman and Essene, 1982) outline fluid evolution paths with increasing grade based on the observed mineral assemblages. These studies describe fluid evolution paths buffered either to CO_2-rich or H_2O-rich compositions, and then, somewhat inexplicably, the fluid composition shifts toward enrichment in the other component. These and similar

ALLEGHENY COLLEGE LIBRARY

observations are probably most simply explained by postulating that the fluid compositions are controlled by the solvus in the H_2O-CO_2-NaCl system rather than solely by the mineral assemblage. Even if the bulk of the fluid is produced at invariant points, as suggested by Greenwood (1975), these will be located at or outside the solvus.

Mercolli (1979) and Walther (1983) describe the metamorphism of quartz nodules and veins in a dolomite matrix in the Campolungo region, Switzerland. Tremolite, calcite, and, locally, phlogopite formed in reaction zones between the quartz nodules and the dolomite. Both authors report abundant fluid inclusions that contain CO_2-H_2O mixtures, some of which have densities that match those expected for fluids present during the metamorphism. They both conclude that the formation of tremolite and calcite from the dolomite–quartz assemblage required an influx of H_2O, and that the tremolite-producing reaction released CO_2 into the fluid resulting in the mixed CO_2-H_2O fluid inclusions. Both authors also report the occasional occurrence of talc in two parageneses: replacing tremolite and as talc–calcite assemblages rather than tremolite–calcite. In his careful analysis of mineral stabilities for the area, Walther concludes that talc and calcite are stable together if the talc is fluorine rich. An alternative interpretation of the observed talc occurrences can be postulated based on the effect of fluid immiscibility on the mineral phase relations. Mercolli (1979) reports a group of ubiquitous highly saline (>30 wt% salts) fluid inclusions trapped along single fractures with a coexisting but immiscible CO_2 fluid (Fig. 4). The phase

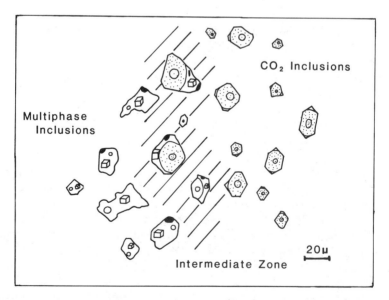

Fig. 4. CO_2 inclusions and multiphase brine inclusions containing halite crystals, coexisting on a single healed fracture. Inclusions of the two immiscible fluids are separated by a narrow intermediate zone in which various proportions of the immiscible phases were trapped (from Mercolli, 1979).

relations for the $MgO-SiO_2-CaO-H_2O-CO_2$ system in the presence of immiscible CO_2-rich and aqueous fluids (Bowers and Helgeson, 1983b) show that talc–calcite and tremolite–calcite assemblages lie on opposite sides of the brine-CO_2 solvus. If the brine is diluted by influx of low-salinity aqueous fluid, the temperature of the consolute point of the CO_2-H_2O solvus is lowered and talc–tremolite assemblages can form, or, as reported in these studies, talc may replace tremolite.

In a study that considers the effect of immiscibility in H_2O-CH_4 fluids, Juster and Brown (1984) suggest that devolatilization reactions in rocks containing unmixed H_2O and CH_4 resulted in assemblages that include coexisting kaolinite and pyrophyllite over a range of conditions. In the presence of pure aqueous fluids and quartz, the kaolinite to pyrophyllite dehydration reaction is invariant; kaolinite and pyrophyllite should not coexist under those conditions. To explain their observations, Juster and Brown (1984) suggest that a CH_4-dominant fluid with low a_{H_2O} triggered kaolinite dehydration reactions to produce pyrophyllite. The resulting high activity of water arrested the reaction until the fluid, once again, became CH_4 rich, repeating the cycle.

Entrapment of Immiscible Fluids

Fluid immiscibility can also be used to explain the occurrence in a number of cases of fluid inclusions with compositions that do not match predicted bulk fluid compositions. For example, fluid inclusions containing brines with salinities of 30–40 wt% dissolved salts are observed associated with separate CO_2 inclusions in calcarous rocks (e.g., Mercolli, 1979; Kreulen, 1980). This association can be readily explained by unmixing of low to moderate salinity H_2O-CO_2-NaCl fluids. Upon fluid phase separation, dissolved salts are strongly partitioned into the aqueous phase. Bowers and Helgeson (1983a) point out that a fluid with $X_{CO_2} = 0.2$ and $X_{NaCl} = 0.04$ may unmix to 14 mol% of a fluid with $X_{CO_2} = 0.004$ and $X_{NaCl} = 0.18$ (~12 molal) coexisting with another fluid with $X_{CO_2} = 0.735$ and $X_{NaCl} = 0.016$.

In some cases, the role of immiscibility in generating saline aqueous and CO_2-rich inclusions is petrographically obvious. Mercolli describes single fractures containing both inclusion types (Fig. 4). The two unmixed fluids may also be found on adjacent fractures (Fig. 5). Although one might expect that trapping immiscible fluids would result in inclusions containing random mixtures of the two fluids, this is not the case. Sterner and Bodnar (1984) prepared fluid inclusions in the laboratory and showed, for three different compositions, that when immiscibility prevails only one or the other of the coexisting fluids was trapped in most cases.

In low- and intermediate-grade schists, pure CH_4, CO_2, or N_2 fluids occur in circumstances in which calculations of C-O-H equilibria predict that water should also be a significant component of the fluid phase (e.g., Mullis, 1975; Hollister and Burruss, 1976; Sisson et al., 1981; Ahrens, 1984). Estimates of

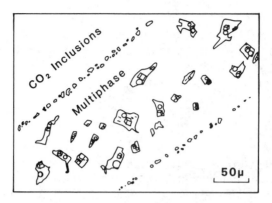

Fig. 5. CO$_2$ inclusions and multiphase brine inclusions containing halite daughter crystals, on separate but closely related healed fractures (from Pecher, 1979).

trapping conditions suggest that these inclusions formed at conditions at which immiscible fluids occur. Commonly, however, the immiscible aqueous phase is rare or absent. Thus, Sisson *et al.* (1981) and Ahrens (1984) observed abundant, almost pure CO$_2$ inclusions trapped along fractures in quartz pods in metamorphosed calc silicates in which the mineral assemblage apparently equilibrated with a fluid phase with a high water fugacity. Similarly, Mullis (1975) reports quartz crystals in hydrothermal fissure veins in the Alps that trapped primary methane inclusions and no water. Presumably, an aqueous fluid was present in both cases to act as a transport medium for the quartz precipitated in the segregations and veins; yet it is not present in the inclusions.

This trapping of only one of the two major components of an unmixed fluid apparently results from physical separation of the immiscible phases prior to trapping. The properties of water molecules are such that water "wets" the walls of the fluid-filled cavity, and the nonaqueous phase is segregated away from the walls. One might assume, therefore, that cracks forming in quartz will trap water selectively as the coexisting nonaqueous fluid would continue to flow along the fracture and leave the system. Alternatively, if the fractures are very narrow, capillary action may "wick" the water out of microcracks into larger fractures, leaving behind globules of the immiscible nonaqueous fluid (CO$_2$, CH$_4$, N$_2$) trapped at irregularities on the fracture surfaces. When such a microcrack subsequently heals, the aqueous phase will have become segregated from the other immiscible fluid. Either process could account for separate trapping of the coexisting immiscible fluids generated by metamorphic reactions. A first step in this process may cause the separation illustrated in Fig. 4 of two immiscible fluids along one fracture. It is likely, if a capillary extraction mechanism operates, that some aqueous fluid does remain, coating the microcrack walls. Molecules of strongly polar compounds, such as water, form an absorbed monomolecular layer on the surface of the fractured grains, and this film of water, at least, will not be removed. Observations by Bodnar (personal communication) support this suggestion; he notes that it is very difficult to trap artificial fluid inclusions along fractures in quartz if there is less than about 5% H_2O to promote crack healing. In fact, the "pure" CO_2 fluid inclusions commonly reported often can be shown to have thin films of water adhering to the inclusion walls (Kreulen, 1980; Sisson *et al.*, 1981). The regular shape of many fluid inclusions and high refractive index contrast between inclusion and host crystal makes it difficult to see less than 15–20 mole fraction H_2O in a CO_2-rich inclusion.

Several other mechanisms for separately trapping unmixed fluid phases can be envisaged. In the case of growing crystals, especially those that grow rapidly, as suggested by Mullis (1976) for Alpine "scepter quartz" grains that contain pure methane inclusions, small globules of immiscible carbonic fluid may become attached to the growing crystal face as "impurities" and are enclosed in the crystal while the aqueous phase that wets the surface remains outside the advancing crystal front. It is also important to remember that, when fluid unmixing occurs, the two unmixed phases differ considerably in density. At most metamorphic conditions, CO_2 is denser than water (Fig. 6). Earlier in this paper we discussed the hypothesis of Walther and Orville (1982) that the contrast between fluid and rock density results in expulsion along microcracks of fluid generated by metamorphic devolatilization reactions. If the fluids are unmixed during this migration, they may become physically separated as a result of density differences as they are expelled upward.

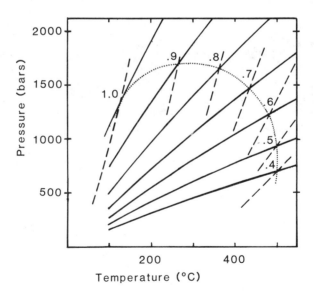

Fig. 6. H_2O (dashed) and CO_2 (solid) isochores labeled with density. The dotted line connects points of equal density. At temperatures and pressures below the dotted line, H_2O is denser than CO_2; the reverse is true everywhere above the dotted line.

At present, we cannot say whether the physical mechanisms discussed above for separating aqueous and carbonic fluids play a significant petrologic role. If two immiscible fluids are present in quantity and can become separated and percolate separately through a pile of metamorphic rocks on a geologically significant time scale, water-absent conditions (Thompson, 1983) may occur during metamorphism.

Postentrapment Separation of Immiscible Fluids

Many CO_2-H_2O inclusions trapped at conditions above the consolute temperature for this system are preserved as inclusions containing unmixed but coexisting fluids at surface conditions. The same rock samples also frequently contain separate, lower density, pure aqueous, and pure carbonic inclusions that could result from redistribution of the fluid trapped in the early generation of inclusions. As mentioned above, unmixing of fluids on the solvus does not, by itself, produce a pure aqueous and a pure carbonic fluid (Fig. 7). However, once a fluid unmixes inside an inclusion, further

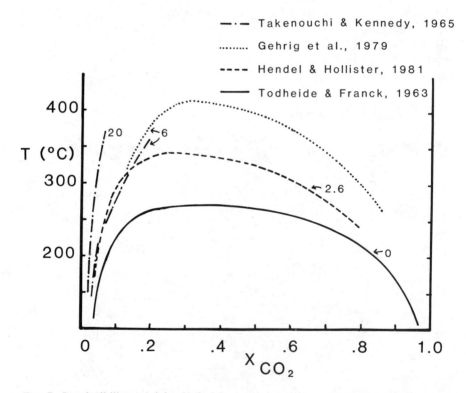

Fig. 7. Immiscibility solvi for H_2O-CO_2 and brine-CO_2 systems. The label on each solvus gives the salinity (wt% NaCl). As shown on the diagram, increasing salinity of the aqueous phase raises and widens the solvus.

phase separation may occur by selective leaking along later microcracks. This is most likely to occur in situations where excess pressure in inclusions is generated during uplift. Because H_2O preferentially wets the walls of inclusions and of fractures, a microcrack extending from an inclusion that contains an unmixed $CO_2 + H_2O$ fluid would preferentially drain the inclusions of H_2O (and possibly also some CO_2) until the pressure in the crack and the inclusion together equaled the confining pressure. Rehealing the fracture could then result in a plane of tiny aqueous inclusions leading away from one large cavity filled with CO_2. Alternatively, most of the water could be drained to the grain boundary leaving only CO_2 in the original inclusion. Considering the time and stress possibilities during unloading and the role played by water in promoting crack propagation, we consider that this process may account for low-density pure CO_2 inclusions in metamorphic rocks.

Discussion

There are, clearly, a large number of possible paths involving immiscibility that a fluid can take from peak metamorphic conditions through sequential episodes of trapping, decrepitation, and rehealing of fractures. It is possible to construct several nonunique explanations to interpret any set of fluid inclusions. We emphasize, however, the fact of fluid immiscibility, that it can occur up to temperatures at least as high as 600°C depending on salt content, and that the required salinities have been observed in low- to medium-grade rocks. At higher temperatures, partial melting can occur. As discussed below, melt + CO_2-rich fluid are expected to coexist in many common situations. Thus, fluid immiscibility should be anticipated over the full range of metamorphic conditions: The least likely temperature range for encountering immiscibility is probably between about 500°C and the beginning of melting. The role of fluid immiscibility in metamorphism was only recognized after serious study of fluid inclusions in metamorphic rocks was underway in the late 1970s and has yet to attract the attention of more than a handful of petrologists studying metamorphic rocks.

Petrologic Applications

Uplift Trajectories

Fluid inclusion studies have been used to constrain postmetamorphic $P-T$ paths of metamorphic terrains, capitalizing on the commonly observed trapping of fluid inclusions of different densities and using the criteria for identifying sequentially trapped inclusions. The basis for these studies is the assumption that after entrapment of an inclusion, its density does not change.

This assumption requires evaluation of the amount of overpressure that can be sustained by each host mineral, as discussed above.

Weisbrod and Poty (1975), Hollister *et al.* (1979), Sisson *et al.* (1981), Hollister (1982), Crawford *et al.* (1984), Selverstone *et al.* (1984), and others have applied the constraint of about 2 kb maximum confining pressure in quartz toward deducing postmetamorphic uplift paths. The argument is that the *P–T* conditions of the host rock must describe a path that passed through or along the isochores for all fluid inclusions found in that rock. The possible segments along these isochores are constrained by the amount of overpressure in the densest fluid inclusion and by the strength of quartz (or other host mineral). This point is illustrated in Fig. 8, which shows isochores for inclusions found in a suite of samples from one metamorphic locality. One isochore (a) corresponds to small, rare, and texturally the earliest formed dense inclusions, and the other two ((b), CO_2 + H_2O) and ((c), CO_2 only) to abundant late-formed low-density inclusions. At some stage during uplift, the rocks had to have crossed all the isochores; if the uplift path went outside the shaded region, the densest inclusions probably would have decrepitated or expanded as a result of excess internal pressures greater than 2 kb.

Entrapment of homogeneous fluid that separates into immiscible phases upon cooling places another constraint on where in *P–T* space an inclusion

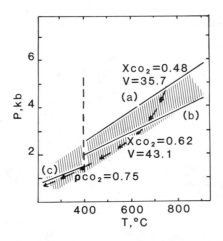

Fig. 8. Three representative isochores for samples from an uplifted regional metamorphic terrane. Isochores (a) X_{CO_2} = 0.48 and (b) X_{CO_2} = 0.62 are for inclusions with mixed H_2O-CO_2 fluids while isochore (c) corresponds to inclusions containing pure CO_2 and therefore presumably trapped below the top of the H_2O-CO_2 solvus (dashed line). The arrows trace a possible uplift path constrained to cross all isochores for inclusions found in the sample and also constrained to lie no further than 2 kb below the isochore for the densest fluid (shaded zone). The latter constraint is imposed by the assumption that >2 kb difference in pressure between the high-density inclusions and ambient conditions would rupture and destroy those inclusions.

was trapped. The consolute point for the CO_2 + H_2O solvus is at 270°C or higher at pressures over 1 kb (Todheide and Franck, 1963), and, as discussed above, the solvus expands substantially with addition of salts. Based on homogenization measurements of inclusions, Hollister *et al.* (1979) concluded that in most metamorphic situations the solvus crest was at a *minimum* of 380°C. Thus, mixed CO_2-H_2O inclusions (isochores (a) and (b), Fig. 8) that can be shown to have been entrapped as a homogeneous phase must have been trapped above this temperature on the isochore for their composition and density. Reference to Fig. 8 illustrates how this constraint plus the occurrence of low-density CO_2 inclusions in the same sample can constrain the uplift and cooling path of a rock. The uplift trajectories deduced with the aid of fluid inclusion studies have shown that many regional metamorphic terranes are characterized by *P–T* paths convex toward the temperature axis. This observation is the basis behind recently published interpretations of regional metamorphism, caused by tectonic processes, which terminates with a rapid uplift phase (Hollister, 1982; Selverstone *et al.*, 1984).

Granulite Facies Fluids

Touret (1971) first reported on the occurrence of dense CO_2 inclusions associated with granulite facies rocks. Subsequently he, his students, and others have noted the nearly ubiquitous association of pure, dense, earliest trapped CO_2 inclusions with granulite facies metamorphism. This observation is strengthened by the fact that granulite xenoliths brought rapidly to the surface from the lower crust contain *only* pure CO_2 inclusions (Bilal and Touret, 1976; Selverstone, 1982). Several workers, notably Collerson and Fryer (1978), Kreulen (1980), Janardhan *et al.* (1979), Friend (1983), Holt and Wightman (1983), and Newton *et al.* (1980), have speculated, based on these observations, that a process of CO_2-streaming from the mantle or the very deep crust not only is the source of the fluid inclusions but also drove the metamorphic reactions by lowering a_{H_2O} as well as by adding additional heat. In the apparent enthusiasm to jump onto the bandwagon of this idea, many of those advocating this model have not taken into account one or more of the following: (1) equilibria of fluid with graphite at high temperature (Hollister and Burruss, 1976; Lamb and Valley, 1984); (2) occurrence of nearly pure, dense CO_2 inclusions in low to medium grades of metamorphism (Crawford *et al.*, 1979; Sisson *et al.*, 1981; Ahrens, 1984); (3) the role of anatectic melt in preferentially extracting H_2O (Fyfe, 1973; Nesbitt, 1980; Powell, 1983a); (4) the occurrence of synmetamorphic CO_2 + H_2O fluid inclusions in at least one granulite facies terrane (Hollister and Burruss, 1976); and (5) the fact that the mechanism of underthrusting of continental crust (Newton *et al.*, 1980) invoked to account for large amounts of CO_2 streaming works against the requirement that these granulites be at higher temperatures than normally expected in the lower crust. This is because the lower portion of the upper crustal slab would be cooled by thermal conduction to the underthrust slab. The upper slab would not subsequently be reheated sufficiently to achieve granulite facies temperatures. None of these objections discount the CO_2 fluid streaming model, but they need to be considered when making petrologic inferences based on fluid inclusion studies. Hansen *et al.* (1984) address several of these issues in their fluid inclusion study of the granulite facies rocks in Karnataka, India, for which they present convincing evidence for local CO_2 infiltration to generate charnockite from biotite gneiss; Newton (this volume) addresses point (1).

The fact is that nearly pure CO_2 fluid inclusions, with densities systematically consistent with independently determined pressures and temperatures of metamorphism, are ubiquitous in most granulite facies terranes. The metamorphic pressures (5–10 kb) suggest lower crustal origins for the granulites; temperatures of $\geq 750°C$ suggest an unusual heat source because steady-state geothermal gradients predict $\sim 500°C$ at 20 km depth; and the existence of continental crust under uplifted and exposed granulite facies terranes demands crustal thickening to bring the granulites to the surface. A convincing case for any specific source for the dense, pure CO_2 inclusions in granulite facies rocks has not yet been made, in our opinion. However, we prefer to assume that CO_2 does not play an active role in the genesis of granulites, based on our own studies, although some arguments for CO_2 influx, at least locally, are convincing (e.g., Hansen *et al.*, 1984). In the following paragraphs, we outline the basis for our preference.

At the temperatures and pressures of granulite facies metamorphism, CO_2 can be produced by one of three mechanisms: (1) oxidation of graphite, (2) reaction of silica- and carbonate-bearing minerals to form calc-silicate minerals plus CO_2, or (3) exsolution from mantle-derived melts. The basis of the first was defined by French (1966) and Eugster and Skippen (1967); its relevance to graphite-bearing granulites has been discussed by Hollister and Burruss (1976), Valley and O'Neil (1981), and Glassley (1983). The second mechanism is suggested by the common occurrences of calc-silicate lithologies in granulite facies terranes. The third was proposed by Touret (1970, 1971) in his first reports on occurrences of pure, dense CO_2 inclusions. He suggested a causal relation between injection of gabbro and the occurrence of CO_2, by expulsion of CO_2 from the gabbro during its crystallization. Textural evidence for this is given by in Selverstone (1982; Fig. 2(c) and p. 31). Solubility of CO_2 in magmas was discussed by Khitarov and Kadik (1973) and Kadik and Lukanin (1973) and demonstrated experimentally by Eggler and Kadik (1979). A theoretical basis for CO_2 solubility in melts is proposed by Spera and Bergman (1982). These studies show that the solubility of CO_2 in magma drops substantially as pressure decreases from upper mantle values to pressures in the lower crust. Therefore, magma of deep origin can exsolve CO_2 when emplaced in the lower crust.

Given one or more of the above as reasonable sources for CO_2 in the lower crust, a process needs to be formulated that selectively removes H_2O from CO_2. Fyfe (1973) suggested that partial melting and removal of granite could effectively dry out the host rock. Olsen (1977) discussed this process in detail in describing migmatite formation in the Baltimore Gneiss. Here, melt formation by anatexis produced a gradient in chemical potential of H_2O away from the site of the first melt, resulting in migration of H_2O and other components down the potential gradient to form the migmatite layering. As the amount of melt increased, it effectively desiccated the adjacent gneiss. Powell (1983b) modeled this process and showed that, with the onset of melting, the mole fraction of CO_2 in vapor coexisting with melt must increase. Based on the experimental data on the partitioning of CO_2 and H_2O between vapor and albite melt given by Eggler and Kadik (1979), and presuming this system adequately models the granite-vapor system, an estimate of the partitioning of CO_2 and H_2O between vapor and melt can be made. The vapor composition is highly pressure sensitive below 5 kb, going from negligible X_{CO_2} at low pressure to a maximum at 5–7 kb and dropping slightly at higher pressures. At 950° and 1050°C, X_{CO_2} of the vapor phase coexisting with the melt exceeds 0.85 at all pressures between 5 and 10 kb. If the melt is removed, the H_2O goes with it, later to be incorporated in the hydrous minerals that crystallize from the melt. All the CO_2-rich fluid need not also have left the rock during this process if it were selectively trapped as fluid inclusions in microcracks in the quartz.

Anatexis involving reaction of hydrous phases such as muscovite or biotite can also result in local formation of pure N_2 inclusions if the mica contains NH_4^+ substituting for K^+. Assuming very low solubility of N_2 in melt analogous to CO_2, the N_2 could enter microcracks and be trapped as fluid inclusions. N_2-bearing fluid inclusions in migmatites have been reported by Touret and Dietvorst (1983).

Extraction of granite melt from granulite facies terranes has not been extensively documented. Arguments suggesting that granite melt extraction is not a viable petrologic process in all granulite facies terrains are reviewed by Newton (this volume). Whether or not granite melts have formed and their possible role in changing the trace element patterns of granulite terranes are, in fact, matters of ongoing controversy. The study of Kenah and Hollister (1983) on anatexis in a granulite facies terrane in British Columbia did establish that a granite composition melt must have formed for which little physical evidence remains in the rocks. By careful study with the microprobe, they located an intergranular rind of orthoclase in some samples that they cited as positive evidence for the existence of the granite melt. Kenah and Hollister (1983) went on to point out that even for situations of $a_{H_2O} < 1$, initial anatexis in deep crustal levels would begin by ~750°C in virtually any rock containing quartz, plagioclase, and biotite. Because these are common constituents of most high-grade metamorphic rocks, at least a small amount of melting should have occurred in virtually all granulite facies terranes, for which metamorphic temperatures over 700°C are invariably reported.

We conclude, therefore, that extraction of H_2O by removal of a granitic melt, leaving a residue with healed fractures containing pure CO_2 or N_2, is a reasonable and viable process for explaining the common association of pure nonaqueous inclusions and granulite facies metamorphism.

Conditions of Vein Formation

The role that fluids play in ongoing petrologic processes is very much a function of the degree to which the fluid phase(s) form an integral part of the evolving rock system or are segregated into fractures and effectively removed from reaction with the solid phases. Well-crafted studies of fluid inclusions that compare compositions and densities of fluids found in inclusions in metamorphic minerals with those in associated vein assemblages can provide data on the source of the vein-forming fluids and the degree to which they are involved in the petrologic evolution of the host rock system.

As discussed above, models of fluid flow through metamorphic rocks suggest that these fluids become channelled into discrete fractures as they move through the rocks (Norris and Henley, 1976; Walther and Orville, 1982; Valley and O'Neil, 1981; Walther and Wood, 1984). The channelways are marked by features that range in scale from trails of fluid inclusions along healed microcracks to veins that may be several meters thick. In the fractures, temperature and pressure changes cause precipitation of materials dissolved in the fluids and generate vein fillings. Of the vein minerals, quartz is the most common at high metamorphic grades, calcite at lower grades consistent with the increase in quartz solubility and decrease in calcite solubility with increasing temperature. In some settings, such as the open cleft veins in the Alps (Poty et al., 1974; Mullis et al., 1983), complex mineral parageneses occur. Many of the fluid inclusion studies in metamorphic rocks concentrate on these vein assemblages because they commonly contain a significantly larger number of fluid inclusions than the adjacent host rocks. In part this probably results from the fact that vein minerals crystallize in a fluid-rich environment. It may also arise if the response to hydraulic fracturing and to deformation of the veins differs from that of the surrounding rock matrix because of the mineralogical differences of the two. Fluids migrating through polymineralic schists and gneisses may selectively open channels along grain boundaries or cleavage planes, whereas in vein quartz aggregates fracturing across grains may be favored. The latter situation appears more favorable to forming and preserving visible fluid inclusions.

As discrete channelways form, these may be localized along zones of weakness that continue to function as conduits for fluid flow. Ramsay (1980) suggests that fractures open, become filled with minerals precipitated from solution, and reopen, often repeatedly. These growing vein minerals may trap fluid inclusions and may, therefore, preserve information on fluid compositions and densities during successive fracture-forming deformation events. Thus, Durney (1972, 1974) has shown that the homogenization temperatures of fluid inclusions in some fibrous vein fillings record temperature changes in the rocks during successive episodes of crack formation.

In many cases, the fluid inclusions in segregations and veins are the same as inclusions in minerals in the host rocks, and the measured compositions and densities agree with those predicted for the metamorphic conditions recorded by the host rocks. These veins apparently are simply conduits for the pervasive metamorphic fluid. However, fluids trapped as inclusions in veins can provide useful information about $P-T$ evolution of the vein system that may not be recorded by the wall rocks. For example, in a detailed study of evolution of fluids in the Alps during retrograde metamorphism Mullis (1983) and Mullis *et al.* (1983) describe fluid compositional changes that require interaction of the fluids with host rock mineral assemblages prior to trapping of the fluids in the vein minerals. They suggest that the fluids were expelled into veins after the retrograde reactions occurred. Those fluids segregated into open fissures during or prior to retrograde reactions could be recognized because they differed from the observed regional pattern of retrograde fluid evolution. These authors also demonstrated a close connection between the subsequent evolution of complex vein mineral parageneses and regional $P-T$ changes.

In low-grade rocks vein assemblages generally form in open fissures (e.g., Mullis, 1979; Saliot *et al.*, 1982; Mullis *et al.*, 1983). As a consequence, the pressure prevailing during fluid inclusion trapping may be lower than lithostatic and, if the fluids are methane rich, may even be lower than hydrostatic. Several methods for determining pressures using fluid inclusions are summarized by Crawford (1981a) and Roedder (1984, chapter 9). Using these techniques Mullis (1976) has documented significant pressure variations during growth of quartz crystals in individual Alpine veins. He suggests that pressure differences of approximately 1000 bars separated successive growth zones. These zones are inferred to represent episodes of vein opening, leading to a pressure decrease, alternating with episodes when the vein sealed. Mullis postulates that vein sealing permitted fluid pressures to build up to close to lithostatic conditions. In a similar study, Saliot *et al.* (1982) suggest that fluid pressures in veins may even have exceeded lithostatic values owing to heating of fluid in sealed vein systems.

Studies of this type on veins can lead to detailed interpretation of interrelated tectonic, thermal, and fluid flow processes in rocks strong enough to sustain open fractures, but also serve to caution those who rely on fluid inclusion studies on vein minerals about extrapolating the data to conditions affecting the host rocks. Studies on veins cutting low-grade metamorphic rocks in environments in which the rocks are strong enough to sustain discrete open fractures suggest that fluids can become segregated and removed from equilibrium with the host rocks. In higher grade schists, hydraulic fracturing is apparently pervasive and the fluids, whatever their source, are commonly in equilibrium with the surrounding mineral assemblages. Fluid inclusion studies can play an important role in examining the transition between these limiting cases. Externally derived fluids, if present, cannot be distinguished from fluids derived by metamorphic devolatilization based on fluid inclusion occurrences alone. Whether they are restricted to vein assemblages or occur on pervasive small-scale fractures is probably a function of the overall geologic setting rather than fluid source. But externally derived fluids may be identified by compositions or densities that cannot be reconciled with compositions and densities expected in the host metamorphic rocks. The most clearcut examples of this are dense, pure water inclusions, commonly identified as late, and probably derived from meteoric water.

Conclusions

Understanding the properties of fluids in the deeper crust requires knowledge of the physical behavior of the fluid phase. The presence of synmetamorphic fluid inclusions even in high-grade, deep-seated rocks supports the conclusion that a discrete fluid phase is present. The occurrence of these inclusions on healed microfractures that cut across grain boundaries suggests that fluid moves through these rocks by hydrofracturing rather than percolating along grain boundaries. A consideration of fluid inclusions trapped in veins suggests that these veins also mark the location of conduits for transfer of metamorphic fluids. During prograde metamorphism, fluids generated by devolatilization reactions have compositions buffered by the enclosing mineral assemblages. Careful analysis of compositions, densities, and textural data suggest that many fluid inclusions in metamorphic rocks represent this synmetamorphic fluid, possibly redistributed from the original trapping sites during uplift, decompression, and cooling and possibly chemically modified by separation of immiscible fluids.

Conditions of entrapment of fluid inclusions in metamorphic rocks, their compositions, and complexities introduced by postentrapment modification are different from those in igneous and hydrothermal rocks. This is due to the long time span for metamorphic processes and to the fact that the fluid phase never forms more than a minor component of most metamorphic systems. In metamorphic rocks, fluid inclusions trapped in growing minerals are rare and observation of several generations of fluid inclusions successively trapped on healed microcracks are common. Consideration of the P–T histories of metamorphic rocks suggests that the earliest preserved fluid inclusions commonly formed near the thermal maximum. Subsequent recrystallization, influx of externally derived fluids, and unmixing of fluids immiscible at low to moderate temperatures may generate new fluid inclusions that can be used to document the evolution of the metamorphic rock as it returns to the surface. Indeed, the application of information from sequentially trapped fluid inclusions has contributed significantly to hypotheses of a period of very rapid uplift following peak metamorphism of many terranes.

Studies of fluid inclusions have documented the variety of fluids present in metamorphic rocks: H_2O, saline brine, CO_2, CH_4, N_2, and mixtures of these. They have raised important petrologic questions, such as the origin of abundant CO_2 in the granulite facies and the origin of nitrogen at all metamorphic grades, and have demonstrated that fluid immiscibility plays a significant petrologic role in metamorphic processes. Finally, fluid inclusions both support and complement mineralogic and geochemical studies in determining the complex evolutionary history of metamorphic terranes.

Acknowledgments

Fluid inclusion research by the authors has been supported by the National Science Foundation, most recently by grants EAR 78-03349 and EAR 81-00398 to M. L. Crawford and by grant EAR 82-17341 to L. S. Hollister. We are indebted to numerous colleagues and students for collaborative efforts and fruitful discussions and to J. Selverstone and J. Walther for careful and constructive reviews of this manuscript.

References

Ahrens, L.J. (1984) A study of fluid inclusions in the Waterville-Vassalboro region, Maine. BA thesis, Princeton University, New Jersey.

Althaus, E., and Johannes, W. (1969) Experimental metamorphism of NaCl-bearing aqueous solutions by reactions with silicates. *Amer. J. Sci.* **267,** 87–98.

Berglund, L., and Touret, J. (1976) Garnet-biotite gneiss in "Systeme du Graphite" (Madagascar): Petrology and fluid inclusions. *Lithos* **9,** 139–148.

Bilal, A., and Touret, J. (1976) Les inclusions fluides des phenocristaux des laves basaltiques du Puy Beaunit (Massif Central Francais). *Bull. Soc. Fr. Mineral. Crist.* **100,** 324–328.

Bischoff, J.L., Radtke, A.S., and Rosenbauer, R.J. (1981) Hydrothermal alteration of graywacke by brine and seawater: role of alteration and chloride complexing on metal solubilization at 200° and 350°C. *Econ. Geol.* **76,** 659–676.

Bowers, T.S., and Helgeson, H.C. (1983a) Calculation of the thermodynamic and geochemical consequences of nonideal mixing in the system H_2O-CO_2-NaCl on phase relations in geologic systems: Equation of state for H_2O-CO_2-NaCl fluids at high pressures and temperatures. *Geochim. Cosmochim. Acta* **47,** 1247–1275.

Bowers, T.S., Helgeson, H.C. (1983b) Calculation of the thermodynamic and geochemical consequences of nonideal mixing in the system H_2O-CO_2-NaCl on phase relations in geologic systems: Metamorphic equilibria at high pressures and temperatures. *Amer. Mineral.* **68,** 1059–1075.

Bowman, J.R., and Essene, E.J. (1982) P–T–$X(CO_2)$ conditions of contact metamorphism in the Black Butte aureole, Elkhorn, Montana. *Amer. J. Sci.* **282,** 311–340.

Burruss, R.C. (1981) Analysis of fluid inclusions: Phase equilibria at constant volume. *Amer. J. Sci.* **281,** 1104–1126.

Burruss, R.C., and Hollister, L.S. (1979) Evidence from fluid inclusions for a paleogeothermal gradient at the geothermal test well site, Los Alamos, New Mexico. *J. Volc. Geothermal. Res.* **5,** 163–177.

Chatterjee, N.D., and Froese, E. (1975) A thermodynamic study of the pseudobinary join muscovite-paragonite in the system $KAlSi_3O_8$-Al_2O_3-SiO_2-H_2O. *Amer. Mineral.* **60,** 985–993.

Collerson, K.D., and Fryer, B.J. (1978) The role of fluids in the formation and subsequent development of early continental crust. *Contrib. Mineral. Petrol.* **67**, 151–167.

Coolen, J.J.M.M.M. (1980) *Chemical Petrology of the Furua Granulite Complex, Southern Tanzania.* Amst Univ Pub (GUA) Ser. 1, vol 13.

Crampon, N. (1973) Metamorphisme hydrothermal en facies salins et penesalins sur l'exemple du complexe salifere de l'extreme Nord Tunisien. *Contrib. Mineral. Petrol.* **39**, 117–140.

Crawford, M.L. (1981a) Fluid inclusions in metamorphic rocks—low and medium grade, in *Mineralogical Association of Canada Short Course Handbook,* edited by L.S. Hollister and M.L. Crawford, **6**, 157–181.

Crawford, M.L. (1981b) Phase equilibria in aqueous fluid inclusions. in *Mineralogical Association of Canada Short Course Handbook,* edited by L.S. Hollister and M.L. Crawford, **6**, 75–100.

Crawford, M.L., Kraus, D.W., and Hollister, L.S. (1979) Petrologic and fluid inclusion study of calc-silicate rocks, Prince Rupert, British Columbia. *Amer. J. Sci.* **279**, 1135–1159.

Crawford, M.L., Stout, M.Z., and Ghent, E.D. (1984) $P-T-X$ (fluid) evolution in Al_2SiO_5-bearing schist, Mica Creek, BC. *Geol. Soc. Amer. Abstr. Progs.* **16**, 478.

Dubessy, J., (1984) Simulation des equilibres chimiques dans le systeme C-O-H, Consequences methodologiques pour les inclusions fluides. *Bull. Mineral.* **107**, 155–168.

Durney, D.W. (1972) Deformation history of the western Helvetic Nappes, Valais, Switzerland. Thesis, London University, England.

Durney, D.W. (1974) Relations entre les temperatures d'homogeneisation d'inclusions fluides et les mineraux metamorphiques dans les nappes helvetiques du Valais. *Bull. Soc. Geol. France* **16**, 269–272.

Eggler, D.H., and Kadik, A.A. (1979) The system $NaAlSi_3O_8$-H_2O-CO_2 to 20 kbar pressure: Compositional and thermodynamic relations of liquids and vapors coexisting with albite. *Amer. Mineral.* **64**, 1036–1048.

Eugster, H.P., and Gunter, W.D. (1981) The compositions of supercritical metamorphic solutions. *Bull. Minéral.* **104**, 817–826.

Eugster, H.P., and Skippen, G.B. (1967) Igneous and metamorphic reactions involving gas equilibria, in *Researches in Geochemistry, 2,* edited by L.S. Abelson, pp. 492–520. Wiley and Sons, New York.

Eugster, H.P., Albee, A.L., Bence, A.E., Thompson, Jr., J.B., and Waldbaum, D.R. (1972) The two phase region and excess mixing properties of paragonite-muscovite crystalline solutions. *J. Petrol.* **13**, 147–179.

Franck, E.U. (1977) Equilibria in aqueous electrolyte systems at high temperatures and pressures, in *Phase Equilibria and Fluid Properties in the Chemical Industry, ACS Symposium Series 60,* Washington.

French, B.M. (1966) Some geological implications of equilibrium between graphite and a C-H-O gas phase at high temperatures and pressures. *Rev. Geophys.* **4**, 223–253.

Frey, M., Bucher, K., Frank, E., and Mullis, J. (1980) Alpine metamorphism along the Geotraverse Basel-Chiasso—a review. *Eclogae. Geol. Helv.* **73**, 527–546.

Friend, C.R.L. (1983) The link between charnockite formation and granite production: Evidence from Kabbaldurga, Karnataka, southern India, in *Migmatites, Melting and Metmaorphism,* edited by M.P. Atherton and C.D. Gribble, pp. 264–276. Shiva Publ. Co., Nantwich.

Frost, B.R. (1979) Mineral equilibria involving mixed volatiles in a C-O-H fluid phase: The stabilities of graphite and silicate. *Amer. J. Sci.* **279**, 1033–1054.

Fyfe, W.S. (1973) The generation of batholiths. *Tectonophysics* **17**, 273–283.

Gehrig, M. (1980) *Phasengleichgewichte und PVT-Daten ternarer Mischungen aus Wasser, Kohlendioxid und Natriumchlorid bix 3 kbar und 550°C.* Thesis, University of Karlsruhe, West Germany.

Gehrig, M., Lentz, H., and Franck, E.U. (1979) Thermodynamic properties of water–carbon dioxide–sodium chloride mixtures at high temperatures and pressures, in *High-Pressure Science and Technology, 1,* Physical properties and material synthesis, edited by K.D. Timmerhaus and M.S. Barber, pp. 539–542.

Ghent, E.D., Knitter, C.C., Raeside, R.P., and Stout, M.Z. (1982) Geothermometry and geobarometry of pelitic rocks, upper kyanite and sillimanite zones, Mica Creek area, British Columbia. *Can. Mineral.* **20**, 295–305.

Glassley, W.E. (1983) The role of CO_2 in the chemical modification of deep continental crust. *Geochim. Cosmochim. Acta* **47**, 597–616.

Granath, V.C.H., Papike, J.J., and Labotka, T.C. (1983) The Notch Peak contact metamorphic member of the Orr Formation. *Geol. Soc. Amer. Bull.* **94**, 889–906.

Greenwood, H.J. (1975) Buffering of pore fluids by metamorphic reactions. *Amer. J. Sci.* **275**, 573–593.

Guilhaumou, N., Dhamelincourt, P., Touray, J.C., and Touret, J. (1981) Etude des inclusions fluides du systeme N_2-CO_2 des dolomites et de quartz de Tunisie septentrionale. Donnees de la microthermometrie et de l'analyse a la microsonde a effet Raman. *Geochim. Cosmochim. Acta* **45**, 657–673.

Hansen, E.C., Newton, R.C., and Janardhan, A.S. (1984) Fluid inclusions in rocks from the amphibolite-facies gneiss to charnokite progression in southern Karnataka, India: Direct evidence concerning the fluids of granulite metamorphism. *J. Metam. Geol.* **2**, 249–264.

Hendel, E.M., and Hollister, L.S. (1981) An empirical solvus for CO_2-H_2O-2.6 wt. % salt. *Geochim. Cosmochim. Acta* **45**, 225–228.

Hollister, L.S. (1973) Immiscible fluid phases during metamorphism, Khtada Lake Area, British Columbia. *Trans. Amer. Geophys. Union* **54**, 1225.

Hollister, L.S. (1982) Metamorphic evidence for rapid (2 mm/yr) uplift of a portion of the Central Gneiss Complex, Coast Mountains, British Columbia. *Can. Mineral.* **20**, 319–332.

Hollister, L.S., and Burruss, R.C. (1976) Phase equilibria in fluid inclusions from the Khtada Lake metamorphic complex. *Geochim. Cosmochim. Acta* **40**, 163–175.

Hollister, L.S., Burruss, R.C., Henry, D.L., and Hendel, E.M. (1979) Physical conditions during uplift of metamorphic terrains, as recorded by fluid inclusions. *Bull. Minéral.* **102**, 555–561.

Hollister, L.S., and Crawford, M.L. (1981) Short course in fluid inclusions: Applications to petrology. *Mineralogical Association of Canada Short Course Handbook,* **6**, 304 pp.

Holloway, J.R. (1981) Compositions and volumes of super-critical fluids in the earth's crust, in *Mineralogical Association of Canada Short Course Handbook,* edited by L.S. Hollister and M.L. Crawford, **6**, 13–38.

Holloway, J.R. (1984) Graphite-CH_4-H_2O-CO_2 equilibria at low-grade metamorphic conditions. *Geology* **12**, 455–458.

Holt, R.W., and Wightman, R.T. (1983) The role of fluids in the development of a granulite facies transition zone in S India. *J. Geol. Soc. London* **140**, 651–656.

Itihara, Y., and Honma, H. (1979) Ammonium in biotite from metamorphic and granitic rocks of Japan. *Geochim. Cosmochim. Acta* **43**, 503–509.

Jacobs, G.K., and Kerrick, D.M. (1981) APL and FORTRAN programs for a new equation of state for H_2O, CO_2, and their mixtures at supercritical conditions. *Computers Geosci.* **7**, 131–143.

Janardhan, A.S., Newton, R.C., and Smith, J.V. (1979) Ancient crustal metamorphism at low P_{H_2O}: Charnockite formation at Kabbaldurga, South India. *Nature* **278**, 511–514.

Juster, T.C., and Brown, P.E. (1984) Fluids in pelitic rocks during very low grade metamorphism. *Geol. Soc. Amer. Abstr. Progs.* **16**, 553.

Kadik, A.A., and Lukanin, O.A. (1973) The solubility-dependent behavior of water and carbon dioxide in magmatic processes. *Geochem. Internat.* **10**, 115–129.

Kenah, C., and Hollister, L.S. (1983) Anatexis in the Central Gneiss Complex, British Columbia, in *Migmatites, Melting and Metamorphism*, edited by M.P. Atherton and C.D. Gribble, pp. 142–162. Shiva Publ. Co., Nantwich.

Kerrich, R. (1976) Some effects of tectonic recrystallization on fluid inclusions in vein quartz. *Contrib. Mineral. Petrol.* **59**, 195–202.

Khitarov, N.I., and Kadik, A.A. (1973) Water and carbon dioxide in magmatic melts and peculiarities of the melting process. *Contrib. Mineral. Petrol.* **41**, 205–215.

Kreulen, R. (1980) CO_2-rich fluids during regional metamorphism on Naxos (Greece): Carbon isotopes and fluid inclusions. *Amer. J. Sci.* **280**, 745–771.

Kreulen, R., and Schuiling, R.D. (1982) N_2-CH_4-CO_2 fluids during formation of the Dome de l'Agout, France. *Geochim. Cosmochim. Acta* **46**, 193–203.

Lamb, W., and Valley, J.W. (1984) Metamorphism of reduced granulites in low-CO_2 vapour-free environment. *Nature* **312**, 56–58.

Lemmlein, G.G. (1956) Formation of fluid inclusions and their use in geological thermometry. *Geochemistry* **6**, 630–642.

Leroy, J. (1979) Contribution a l'etallonage de la pression interne des inclusions fluides lors de leur decrepitation. *Bull. Minéral.* **120**, 584–593.

Luckscheiter, B., and Morteani, G. (1980) Microthermometrical and chemical studies of fluid inclusions in minerals from Alpine veins from the penninic rocks of the central and western Tauern window (Austria/Italy). *Lithos* **13**, 61–77.

Mercolli, I. (1979) *Le inclusioni fluide nei noduli di quarzo dei marmi dolomitici della regione del Campo lungo (Ticino)*. Eid. Techn. Hochschule, Zürich.

Milovskiy, A.V., and Volynets, V.F. (1966) Nitrogen in metamorphic rocks. *Geochem. Internat.* **3**, 752–758.

Mullis, J. (1975) Growth conditions of quartz crystals from Val d'Illiez (Valais, Switzerland). *Schweiz. Min. Pet. Mitt.* **55**, 419–429.

Mullis, J. (1976) Das Wachstumsmilieu der Quarzkristalle im Val d'Illiez (Wallis, Schweiz). *Schweiz. Min. Pet. Mitt.* **56**, 219–268.

Mullis, J. (1979) The system methane-water as a geologic thermometer and barometer from the external part of the Central Alps. *Bull. Minéral.* **102**, 526–536.

Mullis, J. (1983) Evolution and migration of the fluids in the central Alps during the retrograde metamorphism. European Current Res. on Fluid Inclusions Symposium, 6–8 Apr. 1983. Soc. Franc. Mineral. Crist., Orleans, p. 44.

Mullis, J., Dubessy, J., Kosztolanyi, C., and Poty, B. (1983) Fluid evolution in alpine fissures during prograde and retrograde metamorphism along the geotraverse: Lucerne—Bellinzona (Swiss Alps), European Current Res. on Fluid Inclusions Symposium, 6–8 Apr. 1983. Soc. Franc. Mineral. et Crist., Orleans, p. 46.

Naumov, V.B., Balitskiy, V.S., and Khetchikov, L.N. (1966) Correlation of the temperatures of formation, homogenization, and decrepitation of gas-fluid inclusions. *Akad. Nauk SSSR Dokl.* **171(1)**, 146–148.

Naumov, V.B., Khakimov, A.K.H., and Khodakovskiy, I.L. (1974) Solubility of carbon dioxide in concentrated chloride solutions at high temperatures and pressures. *Geochem. Internat.* **11**, 31–41.

Nesbitt, H.W. (1980) Genesis of the New Quebec and Adirondack granulites: evidence for their production by partial melting. *Contrib. Mineral. Petrol.* **72**, 303–310.

Newton, R.C., Windley, B.F., and Smith, J.V. (1980) Carbonic metamorphism, granulites, and crustal growth. *Nature* **288**, 45–50.

Norris, R.J., and Henley, R.W. (1976) Dewatering of a metamorphic pile. *Geology* **4**, 333–336.

Ohmoto, H., and Kerrick, D.M. (1977) Devolatilization equlibria in graphitic systems. *Amer. J. Sci.* **277**, 1013–1044.

Olsen, S.N. (1977) Origin of the Baltimore Gneiss migmatites at Piney Creek, Maryland. *Geol. Soc. Amer. Bull.* **88**, 1089–1101.

Pagel, M. (1975) Cadre geologique des gisements d'uranium de la structure Carswell, Canada. Etude des phases fluides. Thesis 3 cycle, Nancy.

Paxton, S.T. (1984) Occurrence and distribution of ammonium illite in the Pennsylvania, USA coal fields and proposed relationship to thermal maturity. *Geol. Soc. Amer. Abstr. Progs.* **16**, 620.

Pecher, A. (1979) Les inclusions fluides des quartz d'exsudation de la zone MCT himalayen au Nepal Central: donnees sur la phase fluide dans une grande zone de cisaillement crustal. *Bull. Minéral.* **102**, 537–554.

Poty, B., Stalder, H.,and Weisbrod, A. (1974) Fluid inclusions studies in quartz from fissures of western and central Alps. *Schweiz. Min. Pet. Mitt.* **54**, 717–752.

Powell, R. (1983a) Processes in granulite facies metamorphism, in *Migmatites, Melting and Metamorphism,* edited by M.P. Atherton and C.D. Gribble, pp. 127–139. Shiva Publ. Co., Nantwich.

Powell, R. (1983b) Fluids and melting under upper amphibolite facies conditions. *J. Geol. Soc.* **140**, 629–633.

Ramsay, J.G. (1980) The crack-seal mechanism of rock deformation. *Nature* **284**, 135–139.

Rich, R.A. (1979) Fluid inclusion evidence of Silurian evaporites in southeastern Vermont. *Geol. Soc. Amer. Bull.* **90**, 1628–1643.

Roedder, E. (1971) Fluid inclusion studies on the porphyry-type ore deposits at Bingham (Utah), Butte (Montana), Climax (Colorado). *Econ. Geol.* **66**, 98–120.

Roedder, E. (1972) The composition of fluid inclusions, in *Data of Geochemistry.* U.S. Geological Survey Professional Paper **440JJ.**

Roedder, E. (1981a) Origin of fluid inclusions and changes that occur after trapping, in *Mineralogic Association of Canada Short Course Handbook,* edited by L.S. Hollister and M.L. Crawford, **6**, 101–137.

Roedder, E. (1981b) Problems in the use of fluid inclusions to investigate fluid–rock interactions in igneous and metamorphic processes. *Fortschr. Mineral.* **59**, 267–302.

Roedder, E. (1984) Fluid inclusions. *Reviews in Mineralogy 12.* Mineral. Soc. America.

Rosenbusch, H. (1923) *Mikroskopische Physiographie des Petrographisch wichtigen Mineralien.* 5 Auflag von O. Mugge, Stuttgart.

Saliot, P., Grappin, C., Guilhaumou, N., and Touray, J.-C. (1978) Conditions de formation des inclusions fluides hypersalines de quelques quartz de la "zone des gypses" (Vanoise, Alpes de Savoie). *C. R. Acad. Sci. Paris Ser. D* **286**, 379–381.

Saliot, P., Guilhaumou, N., and Barbillat, J. (1982) Les inclusions fluides dans les mineraux du metamorphisme a laumontite-prehnite-pumpellyite des gres du Champsaur (Alpes du Dauphine). Etude du mecanisme de circulation des fluides. *Bull. Minéral.* **105**, 648–657.

Sanford, R.F. (1980) Textures and mechanisms of metamorphic reactions in the Cockeysville Marble near Texas, Maryland. *Amer. Mineral.* **65**, 654–669.

Sauniac, S., and Touret, J. (1983) Petrology and fluid inclusions of a quartz-kyanite segregation in the main thrust zone of the Himalayas. *Lithos* **16**, 35–45.

Schulien, S. (1980) Mg-Fe partitioning between biotite and a supercritical chloride solution. *Contrib. Mineral. Petrol.* **74**, 85–93.

Selverstone, J. (1982) Fluid inclusions as petrogenetic indicators in granulite xenoliths, Pali-Aike volcanic field, southern Chile. *Contrib. Mineral. Petrol.* **79**, 1–9.

Selverstone, J., Spear, F.S. Franz, G., and Morteani, G. (1984) High pressure metamorphism in the SW Tauern window, Austria. *P–T* paths from hornblende-kyanite-staurolite schists. *J. Petrol.* **25**, 501–531.

Shelton, K.L., and Orville, P.M. (1980) Formation of synthetic fluid inclusions in natural quartz. *Amer. Mineral.* **65**, 1233–1236.

Sisson, V.B., Crawford, M.L., and Thompson, P.H. (1981) CO_2-brine immiscibility at high temperatures, evidence from calcareous metasedimentary rocks. *Contrib. Mineral. Petrol.* **78**, 371–378.

Smith, D.L., and Evans, B. (1984) Diffusional crack healing in quartz. *J. Geophys. Res.* **89**, 4125–4135.

Sorby, H.C. (1858) On the microscopical structure of crystals, indicating the origin of minerals and rocks. *Quart. J. Geol. Soc. London* **14**, 453–500.

Spera, F.J., and Bergman, S.C. (1982) Carbon dioxide in igneous petrogenesis. I. Aspects of the dissolution of CO_2 in silicate liquids. *Contrib. Mineral. Petrol.* **74**, 55–66.

Stalder, H.A., and Touray, J.C. (1970) "Fensterquartz" with methane-bearing inclusions from the western part of the northern sedimentary Swiss Alps. *Schweiz. Min. Pet. Mitt.* **50**, 109–130.

Sterner, S.M., and Bodner, R.J. (1984) Synthetic fluid inclusions in natural quartz. I. Compositional types synthesized and applications to experimental geochemistry. *Geochim. Cosmochim.Acta* **48**, 2659–2668.

Stout, M.Z., Crawford, M.L., and Ghent, E.D. (in press) Pressure, temperature and evolution of fluid compositions of Al_2SiO_5-bearing rocks, Mica Creek, B.C., in light of fluid inclusion data and mineral equilibria. *Contrib. Mineral. Petrol.*

Swanenberg, H.E.C. (1980) Phase equilibria in carbonic systems and their application to freezing studies of fluid inclusions. *Contrib. Mineral. Petrol.* **68**, 303–306.

Takenouchi, S., Kennedy, G.C. (1965) The solubility of carbon dioxide in NaCl solutions at high temperatures and pressures. *Amer. J. Sci.* **263**, 445–454.

Tan, T.H., and Kwak, T.A.P. (1979) The measurement of the thermal history around the Grassy Granodiorite, King Island, Tasmania, by use of fluid inclusion data. *J. Geol.* **87**, 43–54.

Thompson, A.B. (1983) Fluid-absent metamorphism. *J. Geol. Soc.* **140**, 533–548.

Todheide, K., and Franck, E.U. (1963) Das Zweiphasengebiet und die kritische Kurve in System Kohlendioxid-Wasser bis zu Drucken von 3,500 bar. *Zeitschr. Chimie. Neufolge.* **37**, 388–401.

Touray, J.-C. (1970) Analyse thermo-optique des familles d'inclusions a depots salins (principalement halite). *Schweiz. Min. Pet. Mitt.* **50**, 67–79.

Touret, J. (1970) Le facies granulite, metamorphisme en milieu carbonique. *C. R. Acad. Sci. Paris Ser. D* **271**, 2228–2231.

Touret, J. (1971) Le facies granulite en Norvege meridionale. II. Les inclusions fluides. *Lithos* **4**, 423–436.

Touret, J. (1974) Facies granulite et fluides carboniques, in *Centenaire de la societe geologique de la Belgique: geologie des domaines cristallins.* Liege, pp. 267–287.

Touret, J. (1977) The significance of fluid inclusions in metamorphic rocks, in *Thermodynamics in Geology,* edited by C.D.G. Fraser, pp. 203–227. D. Reidel, Dordrecht/Boston.

Touret, J. (1981) Fluid inclusions in high grade metamorphic rocks, in *Mineralogical Association of Canada Short Course Handbook,* edited by L.S. Hollister and M.L. Crawford, **6**, 182–208.

Touret, J., and Dietvorst, P. (1983), Fluid inclusions in high-grade anatectic metamorphites. *J. Geol. Soc. London* **140**, 635–649.

Tuttle, O.F. (1949) Structural petrology of planes of liquid inclusions. *J. Geol.* **57**, 331–356.

Valley, J.W., and O'Neil, J.R. (1981) $^{13}C/^{12}C$ exchange between calcite and graphite: A possible thermometer in Grenville marbles. *Geochim. Cosmochim. Acta* **45**, 411–419.

Vidale, R. (1983) Pore solution compositions in a pelitic system at high temperatures, pressures and salinities. *Amer. J. Sci.* **283-A**, 298–313.

Walther, J.V. (1983) Description and interpretation of metasomatic phase relations at high pressures and temperatures. 2. Metasomatic reactions between quartz and dolomite at Campolungo, Switzerland. *Amer. J. Sci.* **283-A**, 459–485.

Walther, J.V., and Orville, P.M. (1982) Volatile production and transport in regional metamorphism. *Contrib. Mineral. Petrol.* **79**, 252–257.

Walther, J.V., and Wood, D.J. (1984) Rate and mechanism in prograde metamorphism. *Contrib. Mineral. Petrol.* **88**, 246–259.

Weisbrod, A., and Poty, B. (1975) Thermodynamics and geochemistry of the hydrothermal evolution of the Mayres pegmatite south-eastern Massif Central (France), Part I. *Petrologie* **1**, 1–16.

Wilkins, R.W.T., and Barkas, J.P. (1978) Fluid inclusions, deformation and recrystallization in granite tectonites. *Contrib. Mineral. Petrol.* **65**, 293–299.

Wlotzka, F. (1961) Geochemistry of nitrogen. *Geochim. Cosmochim. Acta* **24**, 106–154.

Yardley, B.W.D. (1983) Quartz veins and devolatilization during metamorphism. *J. Geol. Soc.* **140**, 657–663.

Yardley, B.W.D., Shepherd, T.J., and Barber, J.P. (1983) Fluid inclusion studies of high-grade rocks from Connemara, Ireland, in *Migmatites, Melting and Metamorphism,* edited by M.P. Atherton and C.D. Gribble, pp. 110–126. Shiva Publ. Co., Nantwich.

Ypma, P.J.M., and Fuzikawa, K. (1980) Fluid inclusion and oxygen isotope studies of the Nabarlek and Jabiluka uranium deposits, Northern Territory, Australia. Proc. Int. Uranium Symp. on the Pine Creek Geosyncline. Int. Atomic Energy Agency, pp. 375–395.

Zirkel, F. (1873) *Die Mikroskopische Beschaffenheit der Mineralien und Gesteine.* Wilhelm Engelmann, Leipzig, 502 pp.

Chapter 2
Fluids of Granulite Facies Metamorphism

R.C. Newton

Introduction

Rocks of the granulite facies of metamorphism occupy a central role in discussions of petrogenesis of the crust. Almost all well-studied examples are Precambrian, which has suggested to many workers that metamorphic temperature regimes operating in the remote geological past were higher than are characteristic of more recent metamorphism. Granulites are widely believed to make up much of the deeper parts of the continents (Fountain and Salisbury, 1981). The depletion of some very high-grade granulites in large-ion lithophile (LIL) elements, such as U, Th, and Rb, relative to normal upper crustal rocks, increases the appeal of the granulite lower crustal model, because of the low heat flow in ancient shield areas (Heier, 1973). The dense minerals pyroxene and garnet, characteristic of quartzofeldspathic granulites, impart elevated densities and seismic velocities, generally appropriate to the lower crust (Smithson and Brown, 1977). Granulite petrogenesis may therefore be fundamental in the accretion and stabilization of the continents.

The actual processes of granulite formation are quite complex and currently under debate. Despite the number of controversies, two facts are now abundantly clear. The first is that granulites are the products of quite specific metamorphic episodes that operated on limited portions of crust over limited periods of time, typically 50–200 m.y. Metamorphic pressure–temperature ($P-T$) regimes recorded in the mineralogy reveal anomalous crustal thickening and heating episodes, which created physical conditions considerably different from the ambient Precambrian geotherms. Commonly, rocks of surficial origin, such as sediments and lava flows, were buried to depths approaching the base of a normally thick continent, about 30 km (Newton, 1983). The rocks undergoing a cycle of metamorphism and subsequent uplift described some path in the $P-T$ plane. Mineralogic and isotopic features were effectively frozen in at some stage in the round trip, often inferred to have been near peak metamorphic $P-T$ conditions. Metamorphic fluids trapped in small amounts by mineral growth have, in some cases, densities consonant with entrapment at high pressures (e.g., Hollister and Burrus, 1976). The second outstanding result of the study of granulites is that they were recrystallized in low-P_{H_2O} environments. Mineral assemblages were formed at temperatures well above hydrous melting in common lithologies. The definitive mineral orthopyroxene formed at water pressures that were near the lower stability limits of its hydrous precursors amphibole and biotite (Phillips, 1980; Valley et al., 1983). In keeping with the dryness of granulite facies metamorphism, fluid inclusions in mineral grains tend strongly to be CO_2 rich and H_2O poor in granulites (Touret, 1981).

Three major mechanisms have been suggested for desiccation of precursor rocks, some of which were initially H_2O rich:

1. Partial melting, with absorption of H_2O into anatectic melts. This is a classic hypothesis that stems from the notable occurrence of migmatites in some transitional regions between amphibolite facies and granulite facies terranes. Quensel (1951) documented several associations of migmatites and charnockites (orthopyroxene gneisses) on four continents.
2. Dilution of initial H_2O with another volatile, most likely CO_2. This process was invoked for subsolidus conversion of amphibole-bearing gneiss to charnockite in southern India by Condie et al. (1982) and Janardhan et al. (1982). The sources of the copious CO_2 needed may have been deep crustal, as from deeply buried sediments, or subcrustal, as in outgassing of a carbonated mantle (Sheraton et al., 1973).
3. Baking out of rocks in shallow contact aureoles prior to high-pressure metamorphism. This is an important possibility suggested by observations of relict contact metamorphism in supracrustal rocks adjacent to the Adirondack anorthosite (Valley and O'Neil, 1984; Valley, 1985). Subsequent deep burial and metamorphism affected volatile-poor rocks. Metamorphism in many places in the Adirondacks was vapor absent or with such small amounts of volatiles that preexisting mineral assemblages buffered the vapor phase.

Any or all of these desiccation mechanisms may have been effective in a given episode of granulite facies metamorphism.

In any of the three low-P_{H_2O} metamorphic processes suggested above, a vapor phase, if it was present, is likely to have been dominated by CO_2, with smaller amounts of other volatiles. Water pressures could not have been higher than a fraction of the total pressures. Study of fluid inclusions has verified this expectation. Since Touret's (1970) discovery that many granulites from around the world contain dominantly CO_2-rich fluid inclusions in the major minerals, other studies have reported similar findings (Coolen, 1981; Hansen et al., 1984a). The present paper reviews some of the most important aspects of the physical chemistry of carbonic fluids at high temperatures and pressures, with emphasis on quantitative deduction of the nature and physical conditions of the metamorphic fluids. The three mechanisms of fluid-deficient metamorphism, desiccation by anatexis, and CO_2 streaming, are shown to have been operative in specific granulite terranes. Which, if any, of the three, was a more fundamental causative factor, remains unresolved, as do other key questions, such as the reason for the antiquity of most granulites and the mechanisms of LIL depletion.

Physical Chemistry of Carbonic Fluids

CO_2 System

CO_2 is, by itself, the most important volatile system for granulite metamorphism. The P–V–T relations have not been fully determined experimentally in the deep-crustal region of interest, namely 6–11 kbar and 600–1000°C, and enough uncertainty in molar volumes under these conditions remains to hamper phase equilibrium calculations at elevated pressure and temperature. The principal experimental data source is Schmulovich and Schmonov (1978), with coverage of the ranges 250–8000 bars and 200–700°C. Various interpolations and extensions of these data have been made. Touret and Bottinga (1979) and Kerrick and Jacobs (1981) used a modified Redlich–Kwong (MRK) equation of state, discussed earlier by Holloway (1977) and Flowers (1979). This equation of state has the form:

$$P = \frac{RT}{V - b} - \frac{a}{\sqrt{T}\, V(V + b)},$$

where the coefficients a and b are fit to the P–V–T data for a given pure gas. In the general form used by Kerrick and Jacobs (1981), the coefficient a was allowed to vary with both temperature and pressure, and b was considered a function of volume in the first term only. A simple and useful polynomial representation of the Schmulovich and Schmonov data was given by Holland (1981).

Fig. 1 shows representative P–V–T data in the two principal ranges of interest. The low-pressure and -temperature region of the subcritical liquid-vapor (LV) equilibrium is of great value in determination of the densities of CO_2 fluid inclusions by microthermometry and is well known from several experimental studies. Lines of constant volume (isochores) intersect the LV line as shown. Microscopic observation with a heating-cooling stage of the temperature of two-phase equilibrium (homogenization temperature) of a CO_2 inclusion uniquely defines its density. If this mini-system has been essentially isochoric and undisturbed since entrapment during metamorphism, the isochore must pass through the metamorphic P–T range, as determined by mineralogic geothermometry-geobarometry. Isochoric behavior

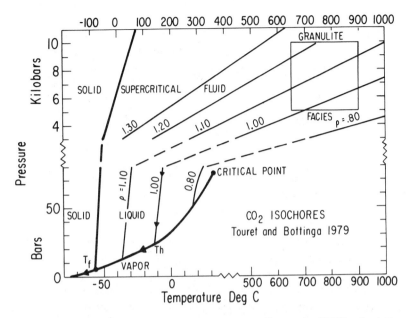

Fig. 1. Phase boundaries of CO_2 and isochores according to the MRK calculations of Touret and Bottinga (1979). Arrows show the refrigeration path of a pure CO_2 mineral inclusion of density 1.00 g/cm³. Densities in this range are compatible with entrapment at granulite facies *T* and *P*.

of fluid inclusions on long time scales is possible only in strong, chemically inert host minerals. Quartz is by far the most useful host mineral.

Pure CO_2 inclusions that become one phase (homogenize) below 25°C have densities greater than about 0.8 g/cm³, and their isochores pass through the granulite facies *P–T* area, compatible with entrapment at peak metamorphic temperatures and pressures. It is a fortunate coincidence that the isochores of greatest interest, those that pass through the typical granulite *P–T* range, intersect the LV line (Fig. 1), so that their densities can be easily determined by microthermometry.

The equilibrium phase behavior of a typical dense liquid CO_2 inclusion upon refrigeration from 25°C is shown in Fig. 1. The internal pressure decreases from several hundred bars at 25°C to a few tens of bars as the specific isochore is traversed. The LV curve is intersected at T_h, the homogenization temperature, and a vapor bubble appears. Further cooling at constant volume takes place along the LV curve until the SLV triple point at −56.6°C is encountered (T_f). When all liquid is frozen, the inclusion cools along the SV curve. In practice, nucleation of vapor and solid may be metastably suppressed, depending on the cooling rate and size of the inclusion. Determination of T_f and T_h are much more reliable in the heating cycle because metastability effects are much smaller.

Melting and homogenization measurements for a given rock sample are commonly collected in histograms. Many granulite samples show T_f very near to the CO_2 triple point, indicating that other components miscible with CO_2 at low temperatures, such as N_2 and CH_4, are present in at most very low concentrations. Homogenization temperatures, T_h, commonly show a considerable spread and may be bimodal in distribution. Fig. 2 shows a typical example of measurements on CO_2 inclusions in quartz. In addition to the two pronounced histogram peaks of Fig. 2, representing groups of inclusions of different density, there is a scattering of measurements to lower temperatures and higher temperatures. The most prominent low-temperature (high-density) peak (region A, Fig. 2) could be interpreted to mark entrapment near peak metamorphic $P-T$ conditions (Hansen *et al.*, 1984a; Rudnick *et al.*, 1984). A low-T_h, high-density tail (region B) has been interpreted as resulting from low-temperature entrapment, either during prograde approach to peak metamorphic conditions, or during nearly isobaric cooling from peak metamorphic conditions (Swanenberg, 1979), as suggested by the arrows in Fig. 2. Sometimes textural evidence of relative chronology of inclusions can decide between such alternatives. The prominent high-T_h, low-density peak (region C) indicates a low-pressure entrapment which, in

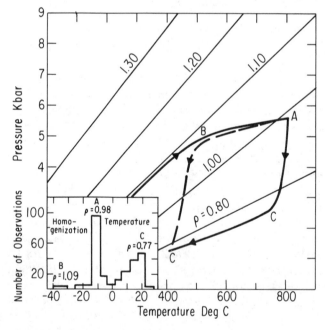

Fig. 2. Typical histogram of CO_2 inclusion homogenization temperatures (inset) from Touret and Dietvorst (1983) and interpretations of $P-T$ path and entrapment episodes. Solid curve with arrows shows postulated metamorphic $P-T$ path, with maximum temperature near 800°C. Dashed line shows alternative return path, with late entrapment of high-density CO_2 during nearly isobaric cooling.

some cases, can be shown petrographically to have occurred relatively late, under post-peak metamorphic $P-T$ conditions (Hollister, 1982; Touret and Dietvorst, 1983). Late entrapment could have taken place during nearly isothermal uplift, perhaps retrapping CO_2 released from decrepitation during unloading, or during step-wise uplifts separated by stages of isobaric cooling, as inferred for the Rogaland granulites of southern Norway (Swanenberg, 1979). It is possible that analysis of CO_2 inclusion density distributions, taken together with similar data for H_2O and other species, will eventually yield more exact information on metamorphic $P-T$ paths. A discussion of the interpretation of $T-P$-time paths from fluid inclusions is given in the article of this volume by Crawford and Hollister.

Other Components in Granulite Fluid Inclusions

CO_2-rich fluid inclusions often contain detectable amounts of CH_4 and/or N_2, components miscible with CO_2 at low temperatures. This is revealed by lowered CO_2 melting points and, recently, by laser-induced Raman spectroscopy of individual fluid inclusions (Rudnick et al., 1984). The triple point of CO_2 is initially lowered by about 1°C per 5 mol% of CH_4. The exact effect of N_2 is not known but is likely to be of the same magnitude. Isochoric homogenization temperatures are also lowered, and three phases (solid, liquid, and vapor) can coexist over a temperature interval in a fluid inclusion.

Burruss (1981) gives an excellent discussion of the physical chemistry of the CO_2-CH_4 system. It is possible, though difficult, to estimate both the composition and bulk density (or isochore) of a CO_2-CH_4 inclusion by visually estimating the relative volume proportions of liquid and vapor phases at final melting, using measured volume data along the three-phase curve. An alternative method, devised by Burruss (1981), is based on accurate measurement of both final melting and homogenization temperatures. The $P-T$ locations of CO_2-CH_4 isochores are not known from experiments at high temperatures and pressures, but MRK calculations have been made (Walther, 1983). For many purposes, the small amounts of CH_4 that are usually present in CO_2-rich inclusions in granulites may be ignored. Neglecting 10 mol% of CH_4 in a CO_2-rich inclusion may result in overestimating the entrapment pressure in a granulite by 1–2 kb if the temperature was 800°C.

Inclusions of nearly pure CH_4 and N_2 have been reported in high-grade metamorphic rocks (Touret and Dietvorst, 1983). These always have low densities and, usually, textural relations indicative of late entrapment. A strange circumstance reported by Touret and Dietvorst is the occurrence of pure N_2 inclusions along with pure CO_2 inclusions in the same healed fracture. These volatiles are miscible and therefore must have been emplaced in separate surges of fluid closely related in time.

CO_2 and H_2O are miscible at granulite facies conditions but nearly completely immiscible at low temperatures, and the presence of NaCl promotes immiscibility to higher temperatures (Bowers and Helgeson, 1983). Crawford and Hollister (this volume) point to the possible role of immiscibility of CO_2 and concentrated brines in generating CO_2-rich fluid inclusions. Discrete inclusions of aqueous brine and mixed CO_2 and H_2O with two or three fluid phases, reported occasionally in granulites, are usually texturally secondary, have low densities, and may be postgranulite facies. Some mixed CO_2-H_2O inclusions in granulites have been shown to be the effect of contamination of original dense CO_2 inclusions by postmetamorphic H_2O introduction (Rudnick et al., 1984). Crystals of the CO_2-H_2O clathrate hydrate, $CO_2\cdot5.75H_2O$, are sometimes observed (Hollister and Burruss, 1976). This compound melts at about 10°C to immiscible CO_2 and H_2O and can be quite useful in detecting the presence of NaCl, which lowers the melting point. H_2O in amounts less than about 20 mol% in CO_2-rich inclusions is very hard to observe microscopically because it forms a surface film on inclusion walls (Roedder, 1972). Hansen et al. (1984a) reported some inclusions from the incipient charnockite zone in southern India in which about 25 mol% of H_2O was visible in capillaries around large irregular CO_2 inclusions, and it may well be that fluid inclusions in some granulites commonly contain up to 20% of unrecognized H_2O.

Immiscibility of CO_2 and H_2O at low temperatures implies some nonideal mixing at granulite facies conditions, and this is an important consideration for phase equilibrium calculations at high temperatures and pressures. Determinations of the activities of H_2O and CO_2 in gas mixtures have been attempted from experimental phase equilibrium data, as from the reaction of calcite and quartz to wollastonite (Greenwood, 1962), from calculations based on the increase of the hydrous melting points of silicate minerals with addition of CO_2 (Bohlen et al., 1982), on direct studies with various experimental techniques, such as the hydrogen sensor cell (Chou and Williams, 1979), and on theoretical equations of state, such as the MRK equation, in which the coefficients a and b can be made functions of composition (Kerrick and Jacobs, 1981). All methods suggest some positive free energy deviations from ideal mixing at several hundred degrees C and several kb, but there is considerable uncertainty in the exact amount. Fig. 3 compares the results of two of the methods. The experimental method of Chou and Williams (1979) predicts increasing nonideality above 8 kb, whereas the MRK calculations suggest that increasing pressure has a smaller effect on nonideality. Calculations based on the experimental albite solidus in the system $NaAlSi_3O_8$-H_2O-CO_2 (Bohlen et al., 1982) give relatively large positive deviations. It is sufficient for purposes of the present discussion to accept the conclusion that some positive nonideality (activities of H_2O higher than mole fractions) exists in the CO_2-rich concentration range.

The physical–chemical properties of CO_2-dominated fluids at high-grade metamorphic conditions can be used, in conjunction with independent estimates of H_2O and O_2 fugacities from mineral equilibria, to define the environments of crystallization of some granulites, as shown in succeeding sections.

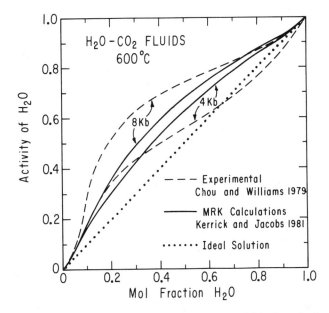

Fig. 3. Activity of H_2O in H_2O-CO_2 fluids at 600°C and 4–8 kb, from hydrogen sensor experimental determination and from MRK calculations. The results show considerable positive deviation from ideal mixing in the low-H_2O range.

H_2O Activities in Granulite Facies Metamorphic Fluids

H_2O-Undersaturated Rock Melting

Coexistence of the assemblage alkali feldspar–plagioclase–quartz is possible at granulite facies *P–T* conditions only with reduced H_2O activity, which prevents melting to granitic liquid. Fig. 4 gives an indication of the a_{H_2O} lowering required. The experimental work of Bohlen *et al.* (1982) on the melting of albite in the system $NaAlSi_3O_8$-H_2O-CO_2 was used as a projection base to derive contours of X_{H_2O} limiting the stability of feldspars + quartz relative to granitic liquid in the system albite–K-feldspar–quartz. If H_2O-CO_2 vapors mix nearly ideally in this *P–T* range, the mole fractions correspond approximately to activities. If the mixing has positive deviations from ideality, as the experimental work and MRK calculations indicate, the activities will be somewhat higher than the mole fractions. It is apparent that the mole fractions of H_2O in a CO_2-H_2O fluid coexisting with feldspars and quartz must be less than about 0.5 over most of the granulite facies range. This is a quantitative expression of the oft-stated opinion that granulite facies metamorphism is attended by low H_2O pressures (for instance, Winkler, 1976).

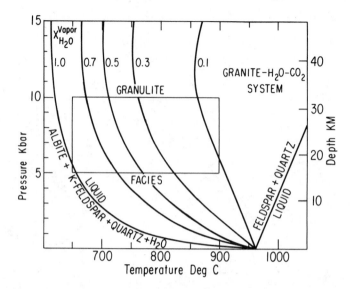

Fig. 4. Solidus curves in the simple granite system $NaAlSi_3O_8$ (albite)-$KAlSi_3O_8$ (K-feldspar)-SiO_2 (quartz)-H_2O-CO_2, projected from experimental data in the system $NaAlSi_3O_8$-H_2O-CO_2 of Bohlen *et al.* (1982). Solidi for $X_{H_2O} = 0$ and $X_{H_2O} = 1$ from Huang and Wyllie (1975). The diagram shows that quartzofeldspathic granulites are stable only at considerably reduced H_2O mole fractions.

If H_2O is not initially very abundant, as in biotite- and amphibole-containing igneous rocks with low porosity, partial melting could absorb H_2O, leaving desiccated residues. This mechanism has been cited for the origin of granulites in various places, such as the Broken Hill area, southern Australia (Phillips, 1980), and Namaqualand, South Africa (Waters, 1984), and is dealt with more fully in the article by Crawford and Hollister in this volume.

Subsolidus Dehydration Equilibria

Amphibole and biotite are stable phases in transitional granulites and some high-grade granulites, where they coexist with pyroxenes, feldspars, and quartz. Calculations of the allowable range of H_2O pressures help to quantify the nature of metamorphic fluids. Stability of pyroxenes and quartz relative to amphibole is limited by the reaction:

$$Ca_2Mg_5Si_8O_{22}(OH)_2 = 2CaMgSi_2O_6 + 3/2Mg_2Si_2O_6 + SiO_2 + H_2O$$

tremolite   diopside   enstatite   quartz   vapor

Valley *et al.* (1983) described the assemblage enstatite–diopside–tremolite–quartz from a metasediment in the Adirondack Mountains. The minerals in their occurrence had close to end-member compositions. They calculated the stability curves of the amphibole–pyroxene equilibrium under conditions of reduced H_2O activity, shown in Fig. 5. It is seen that, in the temperature–pressure range appropriate for the Adirondacks granulites, the H_2O activities were of the order of 0.25–0.50, and these are probably upper limits, since the major impurities in the Valley *et al.* (1983) amphibole are fluorine and aluminum, which would stabilize it to lower a_{H_2O}. This effect was modeled by Valley *et al.* (1983) to yield $a_{H_2O} = 0.11$–0.14. Similar calculations were made by Wells (1979) for the assemblage two pyroxene–quartz–amphibole in granulite gneisses from southwest Greenland. He used the experimental amphibole dehydration curve of Choudhuri and Winkler (1967) and found H_2O activities in the range 0.1–0.3 at P–T conditions similar to those for the Adirondacks. Phillips (1980) found still lower H_2O activities for amphibole–pyroxene equilibrium at Broken Hill.

The limit of biotite stability in many granulites is set by the generalized reaction:

$$KMg_3AlSi_3O_{10}(OH)_2 + 4SiO_2 = 3/2Mg_2Si_2O_6 + KAlSi_3O_8 + H_2O$$

phlogopite   quartz   enstatite   K-feldspar   vapor

The biotite–quartz reaction has been inferred petrographically, as in southern India (Ravindra Kumar *et al.*, 1985). Fig. 6 shows the stability curve of the end-member reaction as determined experimentally. The curve is limited for a pure H_2O vapor by melting at 650 bars and 780°C. Calcula-

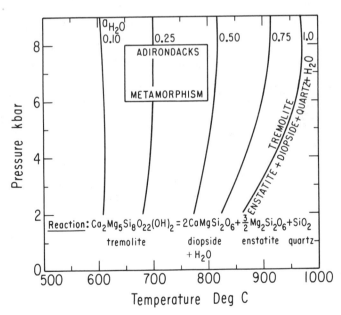

Fig. 5. Stability of the assemblage tremolite–enstatite–diopside–quartz as calculated by Valley *et al.* (1983). The figure shows that the assemblage, observed in the Adirondacks, is stable under the probable metamorphic P–T conditions at low activity of H_2O.

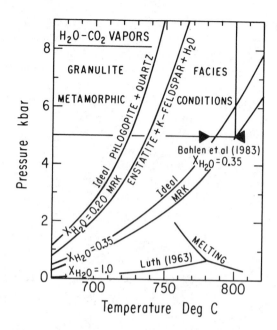

Fig. 6. Stability of the assemblage phlogopite–quartz–enstatite–K-feldspar–vapor in the system K_2O-MgO-Al_2O_3-SiO_2-H_2O-CO_2, calculated from the reaction curve of Luth (1963) for $X_{H_2O} = 1$ with ideal H_2O-CO_2 solution and MRK fluid by Hansen *et al.* (1984a). Either assumption agrees with the reversed experimental determination of Bohlen *et al.* (1983) with $X_{H_2O} = 0.35$. Diagram shows that the mole fraction of H_2O in granulite facies metamorphic fluids coexisting with orthopyroxene and biotite in quartzofeldspathic rocks was less than this value.

tions by Hansen *et al.* (1984a) for ideal H_2O-CO_2 mixing and using an MRK equation of state showed that phlogopite–quartz–enstatite–K-feldspar, a model assemblage for Indian charnockites, could coexist with a vapor under the granulite facies conditions of $T = 700$–$800°C$ and $P = 5$–8 kb only if the mole fraction of H_2O was less than 0.35 during the metamorphism. The calculations for this vapor composition agree with the reversed experimental bracket of Bohlen *et al.* (1983) for the same reaction with CO_2-H_2O vapors, as shown in Fig. 6. Other components that could affect the stabilities of the minerals significantly are Fe^{2+}, Fe^{3+}, Ti, and F. The Fe^{2+} is nearly equipartitioned between biotite and orthopyroxene, and thus will not have a large effect on the equilibrium (Bohlen *et al.*, 1983). The other three components will preferentially stabilize biotite, to a marked extent in the case of Ti (Forbes and Flower, 1974) and F (Westrich, 1981). The firm conclusion can thus be drawn that the calculated stability curves of Fig. 6 represent upper X_{H_2O} limits for the coexistence of orthopyroxene–K-feldspar in the presence of a CO_2-rich vapor phase. Similar calculations on the stability of biotite relative to cordierite and K-feldspar by Phillips (1980) indicated that H_2O activities decreased from 0.5 to 0.3 over the transitional granulite facies zone at Broken Hill.

Although there is disagreement on the amount of nonideality in mixing of H_2O in CO_2-dominated fluids, the consensus is that a_{H_2O} exceeds X_{H_2O} by some factor. Thus, the H_2O activity values yielded by the dehydration equilibria may be thought of as upper limits to the mole fractions of H_2O in the metamorphic fluids.

It should be emphasized that the activity calculations are applicable even in the absence of a metamorphic vapor phase. Thus, if calculated a_{H_2O} and P_{H_2O} were markedly lower than for a pure H_2O fluid ($a_{H_2O} = 1$, $P_{H_2O} = P_{total}$), and if the activities of other volatile species were similarly small, the metamorphism could have been vapor absent, as has been inferred for the crystallization of some Adirondack granulites (Lamb and Valley, 1984).

Volatiles in Cordierite

The composition of cordierite can be expressed by the formula (Fe^{2+}, $Mg)_2Al_4Si_5O_{18} \cdot nH_2O \cdot mCO_2$ for granulite facies rocks where $m + n \leq 1$. Molecular H_2O and CO_2 are contained in structural cavities in the mineral in continuously variable amounts that depend primarily on the partial pressures of these gases during crystallization and on subsequent retentivity. The CO_2/H_2O ratio in Mg–cordierite has been determined experimentally as a function of the corresponding ratio in coexisting vapor at various temperatures and pressures by Johannes and Schreyer (1981). The experimental data were analyzed and expressed in analytic form by Lepezin (1983). Their results indicate that CO_2 is strongly partitioned into vapor, as shown in Fig. 7. Pressure has a small effect on the partitioning, and temperature has almost no effect. There is some evidence that CO_2 tends to block the structural channels and, hence, to promote retention of the original volatile content, during decompression and cooling. It may thus be possible to determine H_2O and CO_2 activities in high-grade metapelites by measuring these volatiles in cordierite. Hörmann et al. (1980) found that X_{H_2O} in cordierite decreased steadily from 0.85 in amphibolite facies rocks to 0.55 in granulite facies rocks in the transitional zone of northern Finland. This would correspond to decrease of X_{H_2O} in a H_2O-CO_2 vapor phase from 0.45 to 0.10 over the transitional interval. These estimates are quite consonant with the H_2O mole fraction critical for granulite facies metamorphism as deduced by the other methods. However, the closing temperatures for CO_2 and H_2O retention in cordierite have not yet been determined for long time scales.

Oxidation Conditions of Granulite Fluids

Granulites range from reduced, graphite-bearing varieties lacking iron oxides, as in paracharnockites of southern India (Ravindra Kumar et al., 1985), to very oxidized, hematite-bearing varieties (Dymek, 1983; Hörmann et al.,

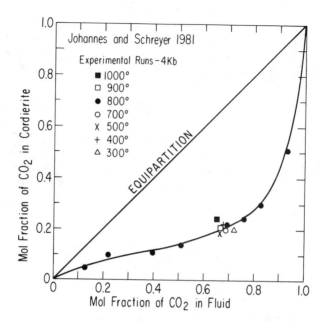

Fig. 7. Experimental partitioning of CO_2 and H_2O between vapor and Mg–cordierite, from Johannes and Schreyer (1981). The partitioning is almost independent of temperature. A cordierite that absorbs fluid of composition 25% CO_2 during crystallization was in equilibrium with a vapor of 75% CO_2.

1980). Analysis of reactions of a fluid phase with Fe-bearing minerals or graphite can further characterize the oxygen fugacity conditions and provides constraints on the nature and amount of the fluid phase present.

Coexisting magnetite and ilmenite provide, in principle, a simultaneous geothermometer and oxygen barometer, if it can be demonstrated that the compositions of the oxide minerals in a high-grade metamorphic rock are representative of peak metamorphic conditions. The reaction of interest is:

$$\underbrace{3FeTiO_3 + 3Fe_2O_3}_{\text{ilmenite solid solution}} = \underbrace{Fe_3O_4 + 3Fe_2TiO_4}_{\text{magnetite solid solution}} + O_2$$

Magnetite rich in Fe_2TiO_4 and ilmenite low in Fe_2O_3 indicate a reduced atmosphere of crystallization. The T–f_{O_2}-composition relations were worked out experimentally by Buddington and Lindsley (1964) at temperatures of 600–1000°C and pressures of 670–2000 bars. Volume considerations indicate that the pressure effect is not likely to be appreciable.

The susceptibility of oxide minerals to exsolution and to retrogressive reaction effectively vitiates their utility for many granulite terranes where metamorphic fluids were abundant, as in southern India (Harris *et al.*, 1982, p. 514). However, for very dry metamorphism, as in granite sheets in the

Scourie terrane of Scotland (Rollinson, 1980) and in the Adirondacks (Bohlen and Essene, 1977), oxide minerals can record previous high-temperature conditions. The latter authors were able to reconstruct original compositions by integrating the bulk compositions of exsolved oxide mineral grains. The inferred oxide mineral temperatures of a number of Adirondack granulites are consistent with reliable mineralogic geothermometers. Calculated f_{O_2} values may thus apply to peak metamorphic P–T conditions.

Another method makes use of orthopyroxene. The basic reaction is:

$$3FeSiO_3 + \tfrac{1}{2}O_2 = Fe_3O_4 + 3SiO_2$$

ferrosilite   magnetite   quartz

$FeSiO_3$ exists in nearly binary solid solution with $MgSiO_3$ (enstatite), and the magnetite contains subsidiary amounts of Fe_2TiO_4. The equation of equilibrium is:

$$-\frac{\Delta G^\circ}{RT} = \ln \frac{a^{Mt}_{Fe_3O_4}}{\left(a^{Opx}_{FeSiO_3}\right)^3 \cdot f^{1/2}_{O_2}},$$

where ΔG° is the difference in free energies of formation of the reactants and the products at variable T and P, f_{O_2} is the oxygen fugacity, and the a's are the activities of the end-member components in the solid solutions. For $FeSiO_3$-rich pyroxenes, the activity may be replaced by the mole fraction, to a considerable degree of validity (Chatillon-Colinet et al., 1983). The composition of magnetite in charnockites is often altered by back reaction, but a self-consistent composition may be found by simultaneous satisfaction of the f_{O_2} condition of Buddington and Lindsley's oxide mineral reaction, in rocks where magnetite and ilmenite are both present. Very Fe-rich pyroxenes coexisting with magnetite and quartz indicate low f_{O_2}, near the quartz–fayalite–magnetite (QFM) equilibrium, whereas intermediate hypersthene requires elevated f_{O_2}.

Fig. 8 shows calculated metamorphic f_{O_2} conditions for two contrasting regimes. Ti-rich magnetite coexisting with ilmenite in Adirondacks metaigneous rocks requires f_{O_2} somewhat below QFM (Lamb and Valley, 1984), which is consistent with Fe-rich orthopyroxenes, whereas the uniformly intermediate hypersthene of acid rocks in southern Karnataka requires f_{O_2} well above QFM. Mineral analyses used in the calculations of the Karnataka data are given in Hansen et al. (1984b).

Lamb and Valley (1984) pointed to a fundamental difference between the two contrasting f_{O_2} regimes of Fig. 8. Basically, high CO_2 pressure in a C-O-H fluid supposes a relatively high oxidation state, which indicates that many of the Adirondacks granulites with reduced oxide minerals were not exposed to CO_2 in high concentration at peak metamorphic conditions. This principle is illustrated by Lamb and Valley's MRK calculations of the relation between H_2O and CO_2 fugacities at metamorphic P–T conditions, shown in Fig. 9. A fluid in the system C-O-H in equilibrium with graphite at a given temperature

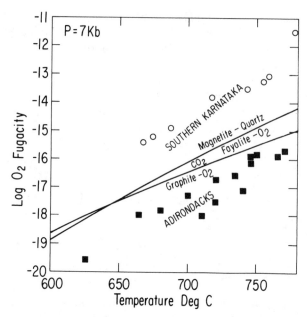

Fig. 8. Apparent oxygen fugacities of crystallization of Adirondacks granulites (filled squares) based on the magnetite–ilmenite method (Lamb and Valley, 1984) and of southern Karnataka charnockites by the method of orthopyroxene–magnetite–quartz (Hansen et al., 1984a). Karnataka paleotemperatures are from the garnet-opx K_D^{Fe-Mg} method of Harley (1984), with mineral analyses from Hansen et al. (1984b). The calculations involve simultaneous satisfaction of the Buddington and Lindsley (1964) magnetite-ilmenite f_{O_2} scale.

and pressure and oxygen fugacity is thermodynamically invariant (has fixed ratios of the vapor species CO_2, H_2O, CH_4, H_2, and CO). At representative conditions of 700°C and 7 kb, the outer solid line in Fig. 9 is the envelope of graphite-vapor coexistence for P_{fluid} equal to P_{total}. For f_{H_2O} and f_{CO_2} to the convex side of this line, a vapor phase may be present, but graphite could not coexist with it.

Contours of f_{O_2} are shown as dashed lines. At a representative Adirondacks f_{O_2} of 10^{-17} bars, the maximum mole fraction of CO_2 possible for $P_{fluid} = P_{total}$ is about 0.3 at point A, on the graphite saturation line. The rest of the vapor is almost entirely H_2O at this point, and this H_2O concentration is too great for orthopyroxene stability. Orthopyroxene may be stabilized at 10^{-17} bars f_{O_2} if f_{H_2O} decreases further, but the log f_{O_2} contour is nearly parallel to the log f_{H_2O} axis, so that f_{CO_2} must remain roughly constant. This means that fluid pressure must decrease well below 7 kb. A vapor phase at pressures markedly lower than rock pressures at high temperatures is not mechanically possible, because of the high ductility of rocks. Apparently, the only way to realize low H_2O fugacities at low f_{O_2} is in the vapor-absent

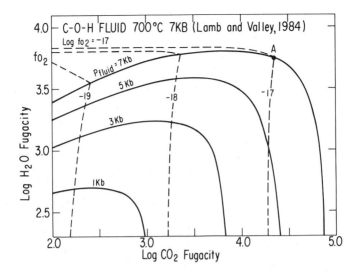

Fig. 9. CO_2-H_2O relations of a C-O-H fluid in equilibrium with graphite at various fluid pressures at 700°C and 7 kb solid pressure, from Lamb and Valley (1984). Contours of log oxygen fugacity are plotted as dashes. For $P_{fluid} = P_{total} = 7$ kb, and a typical Adirondacks f_{O_2} of 10^{-17} bars, the maximum CO_2 of 35 mol% occurs at point A, and the corresponding H_2O content is too high for orthopyroxene stability. Water activities may be lowered at this f_{O_2}, but only at much reduced fluid pressures, which probably indicates vapor-absent metamorphism in many Adirondacks rocks at peak conditions (700°C, 7 kb).

condition, with $P_{fluid} < P_{total}$. Upper limits on P_{fluid} are even more stringent if graphite is not present (activity of graphite less than one), as is true of most of the Adirondacks granulites. These results, which indicate vapor-absent metamorphism in Adirondack charnockites, are further supported by equilibria from a marble xenolith within anorthosite that indicates $P_{H_2O} + P_{CO_2} \ll 7$ kb (Valley, 1985).

Oxidation state is thus seen to set limits to the possible existence and nature of a C-O-H fluid phase in granulite facies metamorphism. Two extremely different granulite facies regimes are exemplified by the southern Karnataka and Adirondacks terranes. Metamorphism in southern Karnataka could have been characterized by active participation of a high-CO_2 vapor phase, with corresponding high f_{O_2}, while metamorphism in much of the Adirondacks took place under reduced, vapor-deficient conditions.

Physical–chemical constraints on the possible existence and nature of participating fluids in granulite facies metamorphism provide a framework for discussion of the origin of granulite terranes. Of special interest are possible mechanisms for desiccation of terranes, for the delivery of carbonic fluids to the site of metamorphism, and for the P–T–fluid histories experienced by the rocks in a metamorphic cycle.

Mechanisms of Granulite Facies Metamorphism

Anatectic Melting

This is a feasible means of absorbing H_2O from a metamorphic site, leaving
desiccated residues, and has been cited as a causative agent of granulite
facies metamorphism in specific terranes, such as the Ivrea Zone of northern
Italy (Schmid, 1978), Namaqualand, South Africa (Waters, 1984), and the
Broken Hill area, southern Australia (Phillips, 1980). Some of the rocks in
the Ivrea Zone have compositions that can be explained as those of common
lithologies after the extraction of varying amounts of a granitic melt. The
"complementariness" of granites and granulites is perhaps the most often-
cited principle in geochemical stratification and stabilization of the crust
(Fyfe, 1973). Further discussion of this mechanism of granulite metamorphism is found in the chapter of this volume by Crawford and Hollister, who
regard it as the most probable explanation of the low-P_{H_2O} regime.

Anatectic melting cannot fully account for a number of important granulite occurrences, however. The isochemical nature of charnockite formation
in the granulite facies transition zone of southern India rules out this mechanism (Condie et al., 1982; Janardhan et al., 1982). Anatexis and melt segregation would have resulted in strong partitioning of some major and minor
elements between the charnockites and their host gneisses, which is not
observed. Another objection is that orthopyroxene-bearing granites and
pegmatites, which would qualify as the anatectic melt fraction, are found in
many granulite terranes. The H_2O that must have been absorbed into these
melts is unaccounted for: some other agency must have removed it after the
melt crystallized. Finally, much quantitative modeling of solid-melt partitioning has resulted in a consensus that Rb depletion patterns are not well
explained by fractionation into and removal by a silicate melt (Condie et al.,
1982; Smalley et al., 1983; Okeke et al., 1983).

Streaming of Low-P_{H_2O} Volatiles

Some rock quarries in southern Karnataka contain veins of coarse charnockite replacing amphibole-bearing gneiss. The cross-cutting textures and isochemical nature of the conversion indicate that orthopyroxene was generated by passage of some kind of low-P_{H_2O} fluid whose access was aided by
deformation. CO_2 is the only vapor species that could be abundant enough to
dilute and carry off H_2O sufficiently to destroy amphibole and biotite. For a
gneiss initially containing 5 wt% of amphibole as the only source of H_2O, the
passage through the rocks of roughly 1 wt% of CO_2 would be required to
produce $X_{H_2O} \sim 0.1$ and so stabilize orthopyroxene. The abundant high-
density CO_2 inclusions in quartz and feldspar of the southern Karnataka
charnockites and their absence in immediately adjacent unconverted
gneisses supports the hypothesis of CO_2 streaming (Hansen et al., 1984a).

Fig. 10 shows the range of P–T isochores of CO_2 inclusions determined by microthermometry of the southern Karnataka rocks. The isochores correspond to the lowest temperature histogram peaks. The isochores pass through the P–T area of granulite facies metamorphism in southern Karnataka, as deduced from mineralogic geothermometry/geobarometry, which is evidence that the CO_2 fluids represent nearpeak metamorphic fluids. Occasionally, some inclusions from the incipient orthopyroxene zone show about 20 mol% of H_2O as a separate phase at room conditions. This is in general accord with experimental and thermodynamic deductions of the threshold H_2O activity for granulite metamorphism. In southern Karnataka, relict igneous textures, exsolution textures such as mesoperthite and spinel trellis texture, and coronal textures are never found, which points to a type of granulite facies metamorphism that is highly fluxed. The elevated oxygen fugacities shown in Fig. 8 for this region are compatible with the presence of a high-pressure CO_2-rich phase during metamorphism. Similar fluid inclusion and mineralogic evidence have been reported for the Furua Complex, Tanzania, granulites by Coolen (1981) and for the West Uusimaa, Finland, granulites by Schreurs (1985).

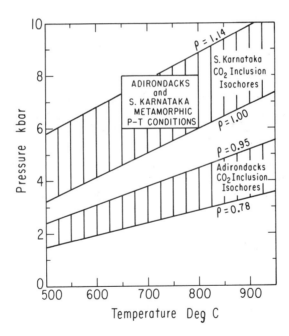

Fig. 10. Isochores of CO_2 inclusions determined by microthermometry of rocks from the Adirondacks (Henry, 1978) and southern Karnataka (Hansen *et al.*, 1984a). The southern Karnataka rocks have densities consonant with entrapment at granulite facies metamorphic pressures, whereas the Adirondacks samples do not. This is further evidence that a CO_2-rich vapor phase coexisted with the southern Karnataka rocks at peak metamorphic pressures but not with the Adirondacks rocks.

Several features of the fluid inclusion evidence remain unexplained. The dense CO_2 inclusions observed by most workers are demonstrably somewhat late in the crystallization sequence: They occupy healed brittle fractures in quartz grains. Thus, many of the fluids observed may be somewhat removed from those present during the initial growth of quartz. Late, low-density CH_4 and N_2 inclusions of the sort observed by Touret and Dietvorst (1983) are enigmatic. Droplets of different fluids that are miscible even at low temperatures are observed as pure phases, sometimes in the same healed fractures. Finally, the sources of the low-P_{H_2O} fluids remain unexplained. CO_2 could have entered the crust from a decarbonating mantle "hot-spot' (Sheraton et al., 1973), or as exhalations from a crystallizing gabbroic underplate (Touret, 1971) or intermediate midcrustal intrusions (Wells, 1979). These mechanisms could also have provided heat for granulite facies metamorphism. Alternatively, continental shelf and shallow marine basin sediments and evaporites, deeply and swiftly buried by tectonic action, might have been a sufficient CO_2 source (Glassley, 1983). Methane and nitrogen seem more likely of biogenic origin than from a mantle source (Kreulen and Schuiling, 1982). However, these volatiles would presumably have been evolved early in a metamorphic sequence, from organic matter or, perhaps, ammoniated clays and micas, rather than late in the sequence, as observed by Touret and Dietvorst.

Vapor-Deficient Metamorphism

It has been shown that reduced oxide minerals and Fe-rich pyroxenes of some granulites, as in the Adirondacks, require that the metamorphism was, in some cases, essentially vapor absent. This inference is reinforced by the observations of the preservation of relict ophitic and rapakivi textures in Adirondack charnockites (Brock, 1980), chilled margins on metagabbros (Gasparik, 1980), and coronitic metagabbros showing weak grain-boundary development of garnet about relict olivine (Johnson and Essene, 1982). Intrusive contacts show extreme variations in apparent f_{CO_2}, f_{O_2}, and oxygen isotope ratios of the sort that would have been produced by shallow contact metamorphism and low-temperature diagenetic processes but that would not have survived pervasive volatile action in granulite metamorphism (Valley et al., 1983; Valley and O'Neil, 1984). If small amounts of interstitial fluid were locally present during Adirondack metamorphism, they were buffered by mineral assemblages to low f_{O_2} and f_{H_2O} and were insufficient to effect equilibration across significant volumes of rock.

The limited amount of fluid inclusion work available for the Adirondack charnockites (Henry, 1978) indicates that most inclusions are relatively low-density CO_2-H_2O mixtures with X_{H_2O} averaging greater than 0.5, which is too water rich to have been in equilibrium with orthopyroxene at peak metamorphic conditions. Fig. 10 shows isochores of the few pure CO_2 inclusions in quartz found by Henry (1978). The inclusions are much too low in density to be isochoric relics of metamorphism at the conditions of 650–800°C and 6–8 kb deduced from the Adirondack granulite metamorphism (Bohlen *et al.*, 1985), but must have been emplaced either early in the metamorphic cycle, or, more probably, during a late uplift stage. The absence of inclusions representative of fluids that could have been in equilibrium with the minerals at peak metamorphic conditions is in keeping with the many other evidences of vapor-deficient metamorphism.

Summary

Many lines of evidence from field observations, petrographic observations, especially of fluid inclusions, and physical–chemical analysis of the fluid inclusions and of mineral assemblages, all point to low H_2O fugacity as the common denominator of granulite metamorphism. The various desiccation mechanisms of absorption of H_2O into anatectic melts, purging of H_2O by copious carbonic fluids, and original dryness, perhaps as the result of low-pressure contact metamorphism prior to high-pressure granulite facies metamorphism, were all effective in specific instances. Which, if any, of these modes were a more fundamental causative factor remains to be determined. The metamorphic histories in given areas, including $P–T$ paths and associated successions of metamorphic fluids, and the mechanisms of LIL depletion, remain as important problems. This information is needed before an assessment of the role of granulite facies metamorphism in the evolution of the continental crust will be possible.

Acknowledgments

The research of the author is supported by National Science Foundation grants EAR 82-19248 and 84-11192. Many stimulating conversations with John Valley guided the course of this paper.

References

Bohlen, S.R., and Essene, E.J. (1977) Feldspar and oxide thermometry of granulites in the Adirondack Highlands. *Contrib. Mineral. Petrol.* **62,** 153–169.
Bohlen, S.R., Boettcher, A.L., and Wall, V.J. (1982) The system albite–H_2O-CO_2: A model for melting and activities of water at high pressures. *Amer. Mineral.* **67,** 451–462.
Bohlen, S.R., Boettcher, A.L., Wall, V.J., and Clemens, J.D. (1983) Stability of phlogopite-quartz and sanidine-quartz: A model for melting in the lower crust. *Contrib. Mineral. Petrol.* **83,** 270–277.
Bohlen, S.R., Valley, J.W., and Essene, E.J. (1985) Metamorphism in the Adirondacks. I. Petrology, pressure and temperature. *J. Petrol.* **26,** 971–992.

Bowers, T.S., and Helgeson, H.C. (1983) Calculation of the thermodynamic and geochemical consequences of nonideal mixing in the system H_2O-CO_2-NaCl on phase relations in geologic systems: Equation of state for H_2O-CO_2-NaCl fluids at high pressures and temperatures. *Geochim. Cosmochim. Acta* **47**, 1247–1275.

Brock, B.S. (1980) Stark complex (Dexter Lake area): Petrology, chemistry, structure, and relation to other green rock complexes and layered gneisses, northern Adirondacks, New York. *Geol. Soc. Amer. Bull.* **91 (Pt. I)**, 93–97.

Buddington, A.F., and Lindsley, D.H. (1964) Iron-titanium oxide minerals and synthetic equivalents. *J. Petrol.* **5**, 310–357.

Burruss, R.C. (1981) Analysis of fluid inclusions: Phase equilibria at constant volume. *Amer. J. Sci.* **281**, 1104–1126.

Chatillon-Colinet, C., Newton, R.C., Perkins, D., and Kleppa, O.J. (1983) Thermochemistry of $(Fe^{2+},Mg)SiO_3$ orthopyroxene. *Geochim. Cosmochim. Acta* **47**, 1597–1603.

Chou, I., and Williams, R.J. (1979) The activity of H_2O in supercritical fluids: H_2O-CO_2 at 600°C and 700°C at elevated pressures. *Lunar Planet. Sci. Conf.* **10**, 201–203.

Choudhuri, A., and Winkler, H.G.F. (1967) Anthophyllit und Hornblende in einegen Metamorphosen Reaktionen. *Contrib. Mineral. Petrol.* **14**, 293–315.

Condie, K.C., Allen, P., and Narayana, B.L. (1982) Geochemistry of the Archean low- to high-grade transition zone, southern India. *Contrib. Mineral. Petrol.* **81**, 157–167.

Coolen, J.J.M.M.M. (1981) Carbonic fluid inclusions in granulites from Tanzania—a comparison of geobarometric methods based on fluid density and mineral chemistry. *Chem. Geol.* **37**, 59–77.

Dymek, R.F. (1983) Fe-Ti oxides in the Malene supracrustals and the occurrence of Nb-rich rutile. *Rapp. Grønlands Geol. Unders.* **112**, 83–94.

Flowers, G.C. (1979) Correction of Holloway's (1977) adaptation of the modified Redlich–Kwong equation of state for calculation of the fugacities of molecular fluids of geologic interest. *Contrib. Mineral. Petrol.* **69**, 315–318.

Forbes, W.C., and Flower, M.F.J. (1974) Phase relations of titan–phlogopite, $K_2Mg_4TiAl_2Si_6O_{20}(OH)_4$: A refractory phase in the upper mantle. *Earth Planet. Sci. Letts.* **22**, 60–66.

Fountain, D.M., and Salisbury, M.H. (1981) Exposed cross-sections through the continental crust: Implications for crustal structure, petrology and evolution. *Earth Planet. Sci. Letts.* **56**, 263–277.

Fyfe, W.S. (1973) The generation of batholiths. *Tectonophysics* **17**, 273–283.

Gasparik, T. (1980) Geology of the Precambrian rocks between Elizabethtown and Mineville, eastern Adirondacks, New York. *Geol. Soc. Amer. Bull.* **91 (Pt. I)**, 78–88.

Glassley, W.E. (1983) Deep crustal carbonates as CO_2 fluid sources: Evidence from metasomatic reaction zones. *Contrib. Mineral. Petrol.* **84**, 15–24.

Greenwood, H.J. (1962) Metamorphic reactions involving two volatile components. *Carnegie Inst. of Washington Yearbook* **61**, 82–85.

Hansen, E.C., Newton, R.C., and Janardhan, A.S. (1984a) Fluid inclusions in rocks from the amphibolite-facies gneiss to charnockite progression in southern Karnataka, India: Direct evidence concerning the fluids of granulite metamorphism. *J. Metam. Geol.* **2**, 249–264.

Hansen, E.C., Newton, R.C., and Janardhan, A.S. (1984b) Pressures, temperatures and metamorphic fluids across an unbroken amphibolite facies to granulite facies transition in southern Karnataka, India, in *Archean Geochemistry,* edited by A. Kröner, A.M. Goodwin, and G.N. Hanson, pp. 161–181. Springer-Verlag, Berlin.

Harley, S.L. (1984) An experimental study of the partitioning of Fe and Mg between garnet and orthopyroxene. *Contrib. Mineral. Petrol.* **86,** 359–373.

Harris, N.B.W., Holt, R.W., and Drury, S.A. (1982) Geobarometry, geothermometry, and late Archean geotherms from the granulite facies terrain of South India. *J. Geol.* **90,** 509–528.

Heier, K.S. (1973) Geochemistry of granulite facies rocks and problems of their origin. *Phil. Trans. Roy. Soc. London* A **273,** 429–442.

Henry, D.L. (1978) A study of metamorphic fluid inclusions in granulite facies rocks of the eastern Adirondacks. Senior Thesis, Princeton University, New Jersey.

Holland, T.J.B. (1981) Thermodynamic analysis of simple mineral systems, in *Thermodynamics of Minerals and Melts,* edited by R.C. Newton, A. Navrotsky, and B.J. Wood, pp. 19–34. Springer-Verlag Berlin/New York.

Hollister, L.S., (1982) Metamorphic evidence for rapid (2 mm/yr) uplift of a portion of the Central Gneiss Complex, Coast Mountains, B.C. *Can. Mineral.* **20,** 319–332.

Hollister, L.S., and Burruss, R.C. (1976) Phase equilibria in fluid inclusions from the Khtada Lake metamorphic complex. *Geochim. Cosmochim. Acta* **40,** 163–175.

Holloway, J.R. (1977) Fugacity and activity of molecular species in supercritical fluids, in *Thermodynamics in Geology,* edited by D.G. Fraser, pp. 161–182. Reidel, Dordrecht.

Hörmann, P.K., Raith, M., Raase, P., Ackermand, D., and Seifert, F. (1980) The granulite complex of Finnish Lappland: Petrology and metamorphic conditions in the Ivalojoki-Inarijarvi area. *Geol. Surv. Finland, Bull.* **308,** 1–95.

Huang, W.-L., and Wyllie, P.J. (1975) Melting reactions in the system $NaAlSi_3O_8$-$KAlSi_3O_8$-SiO_2 to 35 kilobars, dry and with excess water. *J. Geol.* **83,** 737–748.

Janardhan, A.S., Newton, R.C., and Hansen, E.C. (1982) The transformation of amphibolite facies gneiss to charnockite in southern Karnataka and northern Tamil Nadu, India. *Contrib. Mineral. Petrol.* **79,** 130–149.

Johannes, W., and Schreyer, W. (1981) Experimental introduction of CO_2 and H_2O into Mg-cordierite. *Amer. J. Sci.* **281,** 299–317.

Johnson, C.A., and Essene, E.J. (1982) The formation of garnet in olivine-bearing metagabbros from the Adirondacks. *Contrib. Mineral Petrol.* **81,** 240–251.

Kerrick, D.M., and Jacobs, G.K. (1981) A modified Redlich–Kwong equation for H_2O, CO_2, and H_2O-CO_2 mixtures at elevated pressures and temperatures. *Am. J. Sci.* **281,** 735–767.

Kreulen, R., and Schuiling, R.C. (1982) N_2-CH_4-CO_2 fluids during formation of the Dome de l'Agout, France. *Geochim. Cosmochim. Acta* **46,** 193–203.

Lamb, W., and Valley, J.W. (1984) Metamorphism of reduced granulites in low-CO_2, vapour-free environments. *Nature* **312,** 56–58.

Lepezin, G.G. (1983) Determination of the composition of the fluid participating in generation of cordierite-containing rock complexes. *Dokl. Akad. Nauk SSSR* **270,** 132–135.

Luth, W.C. (1963) The system $KAlSiO_4$-Mg_2SiO_4-SiO_2-H_2O from 500 to 3000 bars and 800 degrees. PhD thesis, Pennsylvania State University, Pennsylvania.

Newton, R.C. (1983) Geobarometry of high-grade metamorphic rocks. *Amer. J. Sci.* **283-A,** 1–28.

Okeke, P.O., Borley, G.D., and Watson, J. (1983) A geochemical study of Lewisian

58   R.C. Newton

metasedimentary granulites and gneisses in the Scourie-Laxford area of the northwest Scotland. *Mineral. Mag.* **47**, 1–10.

Phillips, G.N. (1980) Water activity changes across an amphibole–granulite facies
transition, Broken Hill, Australia. *Contrib. Mineral. Petrol.* **75**, 377–386.

Quensel, P. (1951) The charnockite series of the Varberg district on the southwest
coast of Sweden. *Arkiv. Min. Geol.* **1**, 229–332.

Ravindra Kumar, G.R., Srikantappa, C., and Hansen, E.C. (1985) Charnockite formation at Ponmudi in South India. *Nature* **313**, 207–209.

Roedder, E. (1972) The composition of fluid inclusions. U.S. Geological Survey
Professional Paper **440JJ**, 1–164.

Rollinson, H.R. (1980) Iron titanium oxides as an indicator of the role of the fluid
phase during the cooling of granites metamorphosed to granulite trade. *Mineral.
Mag.* **43**, 623–631.

Rudnick, R.L., Ashwal, L.D., and Henry, D.J. (1984) Fluid inclusions in high-grade
gneisses of the Kapuskasing structural zone, Ontario: Metamorphic fluids and
uplift/erosion path. *Contrib.. Mineral. Petrol.* **87**, 399–406.

Schmid, R. (1978) Are the metapelites of the Ivrea-Verbano Zone restites? *Mem.
Sci. Geol.* **33**, 67–69.

Schmulovich, K.I., and Schmonov, V.M. (1978) Tables of thermodynamic properties of gases and liquids (carbon dioxide). Gosdarst. Sluzhba Stardart. Dannykh,
1–l65.

Schreurs, J. (1985) The amphibolite-granulite facies transition in West-Uusimaa,
S.W. Finland. A fluid inclusion study. *J. Metam. Geol.* **2**, 327–342.

Sheraton, J.W., Skinner, A.C., and Tarney, J. (1973) The geochemistry of the Scour-
ian gneisses of the Assynt district, in *The Early Precambrian Rocks of Scotland
and Related Rocks of Greenland,* edited by R.G. Park and J. Tarney, pp. 13–30.
University of Keele.

Smalley, P.C., Field, D., Lamb, R.C., and Clough, P.W.L. (1983) Rare earth, Th-Hf-
Ta and large-ion lithophile element variation in metabasalts from the Proterozoic
amphibolite-granulite transition zone at Arendal, South Norway. *Earth Planet.
Sci. Letts.* **63**, 446–458.

Smithson, S.B., and Brown, S.K. (1977) A model for lower continental crust. *Earth
Planet. Sci. Letts.* **35**, 134–144.

Swanenberg, H.E.C. (1979) Phase equilibria in carbonic systems and their application to freezing studies of fluid inclusions. *Contrib. Mineral. Petrol.* **68**, 303–306.

Touret, J. (1970) Le faciès granulite, métamorphisme en milieu carbonique, *C.R.
Acad. Sci. Paris Ser. D.* **271**, 2228–2231.

Touret, J. (1971) Le faciès granulite en Norwège Méridionale. *Lithos* **4**, 239–249;
423–436.

Touret, J. (1981) Fluid inclusions in high grade metamorphic rocks, in *Short Course
in Fluid Inclusions: Application to Petrology,* edited by L.S. Hollister and M.L.
Crawford, pp. 182–208. Min. Assoc. of Canada.

Touret, J., and Bottinga, Y. (1979) Équation d'état pour le CO_2; application aux
inclusions carboniques. *Bull. Minéral.* **102**, 577–583.

Touret, J., and Dietvorst, P. (1983) Fluid inclusions in high-grade anatectic metamorphites. *J. Geol. Soc. London* **140**, 635–649.

Valley, J.W. (1985) Polymetamorphism in the Adirondacks: Wollastonite at contacts
of shallowly intruded anorthosite, in *The Deep Proterozoic Crust in the N. Atlantic
Provinces,* edited by A.C. Tobi and J. Touret, pp. 217–236. J. Touret. Reidel,
Dordrecht.

Valley, J.W., and O'Neil, J.R. (1984) Fluid heterogeneity during granulite facies metamorphism in the Adirondacks: Stable isotope evidence. *Contrib. Mineral. Petrol.* **85**, 158–173.

Valley, J.W., McLelland, J., Essene, E.J. and Lamb, W. (1983) Metamorphic fluids in the deep crust: Evidence from the Adirondacks. *Nature* **301**, 226–228.

Walther, J.V. (1983) Description and interpretation of metasomatic phase relations at high pressures and temperatures. 2. Metasomatic reactions between quartz and dolomite at Campolungo, Switzerland. *Am. J. Sci.* **283-A**, 459–485.

Waters, D.J. (1984) Dehydration melting and the granulite transition in metapelites from southern Namaqualand, S. Africa. Proc. of the Conf. on Middle to Late Proterozoic Lithosphere Evolution, Cape Town, July 1984.

Wells, P.R.A. (1979) Chemical and thermal evolution of Archaean sialic crust, southern West Greenland. *J. Petrol.* **20**, 187–226.

Westrich, H.R. (1981) F-OH exchange equilibria between mica-amphibole mineral pairs. *Contrib. Mineral. Petrol.* **78**, 318–323.

Winkler, H.G.F. (1976) *Petrogenesis of Metamorphic Rocks* (4th ed.). Springer-Verlag (Heidelberg), 334 pp.

Chapter 3
Reaction Progress: A Monitor of Fluid–Rock Interaction during Metamorphic and Hydrothermal Events

J.M. Ferry

Introduction

When fluid infiltrates a rock and they are not in chemical equilibrium, chemical reactions proceed between fluid and minerals in the rock. Once the stoichiometry of the mineral–fluid reaction and the composition of the fluid is taken into account, the progress of the reaction serves as a quantitative measure of how much fluid the rock chemically interacts with. Reaction progress therefore serves as a natural fossil flux meter for fluid–rock interactions during, for example, metamorphism and hydrothermal events. Numerous applications can be made. On an outcrop scale, reaction progress can determine whether fluid flow was pervasive or was channelized along bedding, fractures, or foliation. On a regional scale, reaction progress can identify metamorphic and hydrothermal infiltration fronts and the relationship between fluid–rock interaction and the degree of metamorphism or alteration. On the scale of an entire metamorphic belt, reaction progress may reveal whether fluid released during metamorphism flows to the surface in a single pass or is recirculated in crustal-scale metamorphic hydrothermal cells.

Reaction progress may be measured by a reaction progress variable (De-Donder, 1920), and the variable has been used extensively in the treatment of nonequilibrium thermodynamics (Fitts, 1962; Prigogine and Defay, 1954). Application of the reaction progress variable to irreversible reactions between minerals and aqueous fluid was pioneered by Helgeson (1968). Helgeson and his students and co-workers developed a rigorous theoretical framework for the use of reaction progress in geochemistry and utilized the variable in numerical simulations of a wide variety of geochemical processes (e.g., Helgeson, 1970; Helgeson et al. 1969, 1970). In a seminal study,

Brimhall (1979) demonstrated that the progress of irreversible mineral–fluid reactions that produced suites of hydrothermally altered granitic rock could be measured from modal data on natural rock samples. Brimhall's work has led, in turn, to the use of the reaction progress variable in studies of mineral–fluid reactions in suites of regional and contact metamorphosed rocks (e.g., Ferry, 1983a; Labotka et al., 1984; Tracy et al., 1983).

The purpose of this contribution is to discuss the various assumptions and techniques by which reaction progress can be used as a quantitative monitor of fluid–rock interaction during the formation of suites of metamorphosed or hydrothermally altered rocks. A number of case studies are reviewed that demonstrate the application of the method. Results of the case studies are discussed in terms of the role of fluid–rock interaction in the petrogenesis of metamorphic rocks.

The Quantitative Measurement of Fluid–Rock Interaction by Reaction Progress: Method and Assumptions

Basic Equations

Reaction progress, ξ, may be formally defined with reference to some arbitrary quantity of rock as:

$$\xi = \Delta n_a / \nu_a, \tag{1}$$

when Δn_a is the change in number of moles of species a in the rock resulting from reaction and ν_a is the stoichiometric coefficient of a in the reaction. The arbitrary quantity of rock may be chosen in terms of mole, mass, or volume units. Volume units, however, are convenient because reaction progress can then be measured in a straightforward way from modal data.

If the arbitrary volume of rock chemically interacts with n_T moles fluid and they are initially out of equilibrium, a mineral–fluid reaction will proceed during which volatile species will either be released to the fluid by the rock, absorbed by the rock from the fluid, or both. The amount of volatile species i must be conserved during reaction, i.e.,

$$X_i^f \left[\sum_j (\nu_j \xi) + n_T \right] = X_i^0 n_T + \nu_i \xi, \tag{2a}$$

where X_i is the mole fraction of species i in the fluid, superscript 0 refers to conditions before fluid–rock reaction, superscript f refers to conditions after fluid–rock reaction, and the sum is taken over all volatile species j, including i, that participate in the mineral–fluid reaction. For a rock that experiences not one but k linearly independent mineral–fluid reactions as it chemically interacts with n_T moles fluid,

$$X_i^f \left\{ \sum_k \left[\sum_j (v_{j,k} \xi_k) \right] + n_T \right\} = X_i^0 n_T + \sum_k (v_{i,k} \xi_k), \tag{3}$$

where ξ_k is the progress of the kth reaction and $v_{j,k}$ and $v_{i,k}$ are the stoichiometric coefficients of each volatile species j, and volatile species i in particular, that participates in the kth reaction. Eqs. (2a) and (3) simply state that the number of moles of volatile species i before reaction plus the number of moles of species i produced or consumed by the reaction is equal to the number of moles of species i after reaction.

Eqs. (2a) and (3) define the formal link between the progress of mineral–fluid reaction(s) and the amount of fluid with which an arbitrary amount of rock interacts during the reaction. Further, the equations identify the information that is required to use reaction progress as a quantitative monitor of how much fluid rocks have chemically interacted with: (1) determination of mineral–fluid reactions that occurred (various v terms); (2) measurement of the progress of the reactions (ξ terms); (3) the composition of fluid before mineral–fluid reaction (X_i^0); and (4) the composition of fluid after mineral–fluid reaction (X_i^f). Techniques for determining or estimating the various v, ξ, and X_i terms in Eqs. (2a) and (3) from suites of metamorphosed or hydrothermally altered rocks are summarized in the next four sections. Once numerical values for the v, ξ, and X_i terms have been substituted into Eq. (2a) or (3), the only remaining unknown is n_T, which then may be readily solved for.

Formulation of Mineral–Fluid Reactions during Metamorphism and Hydrothermal Alteration

Petrologists commonly model mineral–fluid reactions in metamorphosed or hydrothermally altered rocks as mass balance relations among the compositions of mineral and fluid species based on a subjective appraisal of which minerals were created and which destroyed during the metamorphic or hydrothermal event. While only approximate, this method is sometimes adequate, especially in rocks that contain few minerals with simple mineral chemistries. A rigorous, general strategy for systematically determining the possible reactions that have occurred during the formation of a suite of metamorphosed or hydrothermally altered rocks, however, has been presented by Thompson (1982a, 1982b). Following his method, a set of linearly independent system components is chosen that satisfactorily describes the chemical composition of the fluid–rock system of interest as a whole. Let c_s be the number of system components. Next, a set of linearly independent phase components is chosen that satisfactorily describe the chemical composition of each mineral phase in the suite of rock samples and of the fluid phase. Following Thompson's (1982a) suggestion, it is useful to describe the composition of the mineral solid solutions with a single additive component (usually the formula for a conventional end-member mineral component) and one or more exchange components. Let c_p be the number of phase components in the fluid–rock system.

If the system and phase components are linearly independent, $c_p \geq c_s$. The difference, $n_r = c_p - c_s$, is the number of linearly independent reaction relationships that can be written among the phase components, and these reaction relationships represent one description of all the possible ways in which mass may be transferred among the minerals and fluid. Many other equivalent descriptions are possible if different but equivalent choices of phase components are made. Of the n_r reactions, there are n_{ex} exchange reactions, involving exchange components only. The exchange reactions can always be written down by inspection. The remaining $n_{nt} = n_r - n_{ex}$ reactions are net-transfer reactions that involve at least two additive components and may or may not involve exchange components as well. The exchange reactions neither create nor destroy minerals; they are only capable of changing mineral composition. The net-transfer reactions control the creation and destruction of minerals and fluid species. The net-transfer reactions may also change mineral composition. The coefficients on the volatile species in the net-transfer reactions supply the necessary numerical values of v_i, v_j, $v_{i,k}$, and $v_{j,k}$ in Eqs. (2a) and (3).

The reaction relationships among the phase components are only possible reactions that may occur among the minerals and fluid. The reactions that actually occurred during metamorphism or hydrothermal alteration must be determined by measuring the progress of the n_r reactions from data on rock suites. For the purpose of using reaction progress as a monitor of fluid–rock interaction, progress of the exchange reactions is irrelevant because they do not cause a net production or destruction of components of the fluid phase (cf. Eqs. (2a) and (3)). The only reactions whose progress must be measured are the net-transfer reactions.

Measurement of the Progress of Net-Transfer Reactions

Modal Abundance Method

From Eq. (1), a change in number of moles of mineral species a in a rock as a result of reaction, and hence ξ, may be monitored by a change in the species' modal abundance. Mineral species a may represent either a mineral of fixed composition or a component in a mineral solid solution. If the reaction involves no change in volume of the condensed mineral phases,

$$\xi = (V_a^f - V_a^0)/(\overline{V}_a v_a), \qquad (4)$$

where $\overline{V}_a$ is the molar volume of mineral species a and V_a is the volume of a per arbitrary reference volume of rock. The volume amount of a before reaction may be estimated by examining unmetamorphosed or unaltered rocks. The volume amount of a after reaction may be estimated by examining metamorphosed or altered rocks in which the reaction has progressed.

Most mineral–fluid reactions involve a change in volume of the mineral phases, and in this case,

$$\tilde{V}_a^f = \frac{\bar{V}_a^0 V^0 + \bar{V}_a \nu_a \xi}{V^0 + \Delta\bar{V}_s \xi},$$ (5a)

where $\Delta\bar{V}_s$ is the volume of reaction of the condensed phases, $\tilde{V}_a$ is the volume fraction (mode) of a, and V^0 is the arbitrary reference volume of rock before reaction. If an arbitrary reference volume of rock after reaction, V^f, is chosen, the equation corresponding to (5a) is:

$$\tilde{V}_a^f = \frac{\bar{V}_a^0(V^f - \Delta\bar{V}_s \xi) + \bar{V}_a \nu_a \xi}{V^f}.$$ (5b)

Eq. (5b) is particularly useful if rocks of interest contained no species a before reaction ($\bar{V}_a^0 = 0$); then,

$$\xi = V_a^f/(\bar{V}_a \nu_a).$$ (5c)

For a suite of rocks in which there has been progress of k linearly independent reactions, the progress of all the k reactions is related to the modal abundance of mineral species a according to:

$$\tilde{V}_a^f = \frac{\bar{V}_a^0 V^0 + \sum_k (\bar{V}_a \nu_{a,k} \xi_k)}{V^0 + \sum_k (\Delta\bar{V}_{s,k} \xi_k)},$$ (6a)

or

$$\tilde{V}_a^f = \frac{\bar{V}_a^0 \left[V^f - \sum_k (\Delta\bar{V}_{s,k} \xi_k) \right] + \sum_k (\bar{V}_a \nu_{a,k} \xi_k)}{V^f},$$ (6b)

depending on the reference quantity of rock and where $\Delta\bar{V}_{s,k}$ is the volume of reaction for the condensed phases of the kth reaction. To obtain the full set of ξ_k values, k linearly independent equations such as (6) must be solved simultaneously, which, in turn, requires modal and molar volume data on k different mineral species. Ferry (1980), Rumble *et al.* (1982), and Tracy *et al.* (1983) all used the modal abundance method to evaluate reaction progress in metamorphosed impure carbonate rocks.

Phase Ratio Method

In general, progress of a mineral–fluid reaction changes the ratio of minerals present, and ξ therefore may be monitored by a change in the molar ratio of two mineral species in a rock caused by the reaction:

$$(n_a^f/n_b^f) = (n_a^0 + \nu_a \xi)/(n_b^0 + \nu_b \xi),$$ (7)

where a and b are the two mineral species and n_a and n_b are moles of the two mineral species per reference quantity of rock. Eq. (7) does not monitor ξ if $n_a^0 = n_b^0 = 0$, i.e., if the rocks under consideration contained neither species a nor species b before reaction. For a suite of rocks in which there has been progress of k linearly independent reactions, the progress of all the k reactions is related to the change in molar ratio of mineral species a and b by:

$$\frac{n_a^f}{n_b^f} = \frac{n_a^0 + \sum_k (\nu_{a,k} \xi_k)}{n_b^0 + \sum_k (\nu_{b,k} \xi_k)},$$ (8)

where $\nu_{a,k}$ and $\nu_{b,k}$ are the stoichiometric coefficients of species a and b, respectively, in the kth reaction. To obtain the full set of ξ_k values, k linearly independent equations such as (8) must be solved simultaneously, which, in turn, requires modal and molar volume data on $k + 1$ different mineral species. The practical advantage of Eqs. (7) and (8) over Eqs. (5) and (6) is that $\Delta \overline{V}_s$ terms for the net-transfer reactions need not be known in order to obtain numerical values of reaction progress. Ferry (1983a) used the phase ratio method to evaluate reaction progress in a suite of progressively metamorphosed impure carbonate rocks.

Phase Composition Method

In general, progress of a mineral–fluid reaction changes the composition of mineral solid solutions in a rock, and ξ therefore may be monitored by the change in composition of mineral solid solutions caused by the reaction:

$$X_{c,m}^f \left(n_m^0 + \sum_d \nu_d \xi \right) = X_{c,m}^0 n_m^0 + \nu_c \xi,$$ (9)

where $X_{c,m}$ is the mole fraction of component c in mineral solid solution m, n_m is the number of moles mineral m per reference quantity of rock, and the sum is taken over all d components (including c) that are contained in solid solution m. If Thompson's (1982a, 1982b) scheme for representing mineral composition is utilized, however, species d refers *only* to the single additive component of mineral solid solution m. Eq (9) does not monitor ξ if $X_{c,m}^0 = 1$ and $\nu_{d \neq c} = 0$, i.e., if mineral m is effectively not a solid solution. For a suite of rocks in which there has been progress of k linearly independent reactions, the relationship between the progress of the reactions and the change in mineral composition caused by reaction is:

$$X_{c,m}^f \left\{ n_m^0 + \sum_k \left[\sum_d (\nu_{d,k} \xi_k) \right] \right\} = X_{c,m}^0 n_m^0 + \sum_k (\nu_{c,k} \xi_k).$$ (10)

To obtain the full set of ξ_k values, k linearly independent equations such as (10) must be solved simultaneously, which, in turn, requires composition data for k linearly independent phase components in solid solution. The practical advantage of Eqs. (9) and (10) over Eqs. (5)–(8) is that reaction progress may be measured without any modal data for altered or metamorphosed rocks. Ferry (1983a, 1984a), Labotka *et al.* (1984), Thompson *et al.* (1982), and Tracy *et al.* (1983) used the phase composition method to evaluate reaction progress both in suites of metamorphic rocks and in the products of hydrothermal phase equilibrium experiments.

Discussion

Measurement of reaction progress by the modal abundance, phase ratio, and phase composition methods are not mutually exclusive. They are simply the basis for linear equations in ξ that can be solved in any combination. Both Ferry (1983a) and Thompson *et al.* (1982), for example, used combinations of the three methods in their analyses of reaction progress during metamorphism.

Determination of reaction progress involves collecting modal and/or mineral composition data for metamorphosed or hydrothermally altered rocks (variables with superscript f in Eqs. (4)–(10)) and comparing them with modal and mineral composition data for unmetamorphosed or unaltered rocks (variables with superscript 0 in Eqs. (4)–(10)). Erroneous results can be obtained if the unmetamorphosed or unaltered rocks do not actually represent the true protoliths. If the altered or metamorphosed rocks initially had compositions different from the unaltered or unmetamorphosed rocks chosen as protoliths, then these initial compositional differences will be incorrectly interpreted by Eqs. (4)–(10) as differences caused by mineral–fluid reaction. Reaction progress can be determined rigorously from Eqs. (4)–(10) only if the composition and mineralogy of the protolith of all altered or metamorphosed rocks of interest can be determined. In the case of hydrothermally altered plutonic rocks, samples of unaltered igneous rocks, if homogeneous, represent an obvious choice. For example, the composition and mineralogy of unaltered Butte quartz monzonite are uniform (Brimhall, 1979). In his study of hydrothermal alteration during formation of the porphyry copper deposit at Butte, Brimhall measured reaction progress by assuming all altered rocks were derived from a quartz monzonite protolith. In metamorphic terranes, however, lithologies are commonly heterogeneous. If metamorphic reactions are isochemical, the protolith of a particular high-grade metamorphic rock of interest may be reconstructed as a collection of minerals observed in lower grade rocks in proportions that give the bulk chemical composition of the high-grade rock. Ferry (1984b) used this approach to reconstruct the low-grade protoliths of a suite of pelitic schists. Reaction progress may also be measured by modeling metamorphism of a suite of heterogeneous rocks in terms of the average rock at each metamorphic grade. Eqs. (4)–(10) are then applied, assuming that rocks at high metamorphic grade evolved from rocks with properties the average of those exposed at lower metamorphic grade. Ferry (1983a) utilized the method to calculate reaction progress during progressive metamorphism of a heterogeneous suite of impure carbonate rocks. Both approaches to metamorphic rocks could be invalid in Barrovian metamorphic terranes where high-grade rocks may not have evolved from rocks with properties now exposed at low grades (England and Richardson, 1977; Spear and Selverstone, 1983). The approaches are probably satisfactory, however, in Buchan and contact metamorphic terranes where high-grade rocks do apparently evolve from rocks such as those exposed at lower grades (Spear, 1984).

Reaction progress can be measured only either if one or more elements are conserved in the mineral phases during reaction or if the reaction conserves volume. For a given quantity of rock after reaction, conservation of an element (or elements) or volume allows the equivalent amount of rock before reaction to be determined. If mineral–fluid reaction conserves neither any elements nor volume (e.g., if bulk dissolution of rock occurs), then equivalent quantities of rock before and after reaction can never be determined and Eqs. (5)–(10) cannot be applied.

Estimation of Fluid Composition after Mineral–Fluid Reaction

Bird and Norton (1981) showed that minerals and fluid from the Salton Sea geothermal field are close to chemical equilibrium. Mineral–fluid reaction during hydrothermal events probably occurs on a time scale similar to those in a geothermal field while reaction during metamorphism occurs on a much longer time scale. Data from the Salton Sea geothermal field therefore may be taken as evidence that reactions proceed during fluid–rock interaction until equilibrium is closely approached in most instances of metamorphism and hydrothermal alteration. Walther and Wood (1984) suggest that mineral–fluid equilibrium should be attained almost instantaneously on a geologic time scale in metamorphic and hydrothermal environments. The composition of fluid after mineral–fluid reaction (X_i^f, Eqs. (2a) and (3)), therefore, can be calculated as that of fluid in equilibrium with altered or metamorphosed rocks at the P–T conditions of the hydrothermal or metamorphic event. The calculation of fluid composition from mineral–fluid equilibria involves solving sets of simultaneous energy balance equations as reviewed, for example, by Ferry and Burt (1982). All published studies that have used reaction progress as a monitor of fluid–rock interaction have assumed that the composition of fluid after mineral–fluid reaction is that of fluid in chemical equilibrium with the altered or metamorphosed rock (e.g., Ferry, 1980, 1981, 1983a, 1983b, 1983c, 1984a, 1984b; Labotka et al., 1984; Rumble et al., 1982; Tracy et al., 1983). This is a conservative action because it results in minimum estimates of n_T from Eqs. (2a) and (3). If mineral–fluid equilibrium is not attained or if fluids flow through rock and do not chemically react with it at all, then amounts of fluid involved in fluid–rock interaction are greater than amounts calculated from Eqs. (2a) and (3).

Estimation of Fluid Composition before Mineral–Fluid Reaction

The composition of fluid before mineral–fluid reaction (X_i^0 Eqs. (2a) and (3)) is the most difficult variable in the equations to evaluate. Some guidance may be offered by a comparison between the composition of fluid with which

altered or metamorphosed rocks were in chemical equilibrium after mineral–
fluid reaction and the composition of volatiles produced by the reaction. For
example, metamorphosed carbonate rocks commonly record equilibrium
with H_2O-rich fluids while the reactions by which the rocks developed
evolved a CO_2-rich mixture of volatiles (Ferry, 1983b; Rice and Ferry, 1982).
The fluid with which these carbonate rocks chemically interacted must have
been nearly pure H_2O. Sulfide-rich schists studied by Ferry (1981) evolved
an H_2S-rich mixture of H_2S and CO_2, yet were in equilibrium during metamorphism with approximately binary CH_4-H_2O fluid solutions. The fluid
with which the sulfidic schists chemically interacted must have been a nearly
pure CH_4-H_2O mixture. Additional guidance may be offered by the stoichiometry of the mineral–fluid reaction that proceeded during fluid–rock interaction. Hydration reactions require that rocks interacted with H_2O-bearing
fluids; carbonation reactions require that rocks interacted with CO_2-bearing
fluids, etc. Finally, stable isotope data on metamorphosed or hydrothermally
altered rocks may suggest possible compositions of fluids before mineral–
fluid reaction. For altered rocks that have an isotopic signature of interaction
with meteoric water, for example, the chemical composition of fluid with
which they interacted was likely pure H_2O or nearly so. In practice, the most
conservative strategy is to choose X_i^0 as far in composition from X_i^f as is
permitted by the available constraints because the larger the difference between X_i^0 and X_i^f, the smaller the calculated value of n_T. The strategy leads to
minimum estimates of n_T calculated from Eqs. (2a) and (3) and is therefore
consistent with the consequences of estimating the composition of fluid after
mineral–fluid reaction as that of fluid with which altered or metamorphosed
rocks were in chemical equilibrium.

The Quantitative Measurement of Fluid–Rock Interaction by Reaction Progress: Case Studies

General Statement

Eq. (2a) may be explicitly solved for n_T:

$$n_T = \xi \frac{\nu_i - X_i^f\left(\sum_j \nu_j\right)}{X_i^f - X_i^0}.$$ (2b)

If mineral–fluid reaction occurs in a rock (i.e., $\xi > 0$), reaction proceeds
without fluid–rock interaction (i.e., $n_T = 0$) if and only if the reaction evolves
a mixture of volatiles with composition exactly the same as the composition
of fluid with which the rock is in chemical equilibrium (i.e., $X_i^f = \nu_i/\sum_j\nu_j$). In
all other instances, fluid–rock interaction must accompany reaction, and Eq.
(2b) provides an estimate of how much fluid is involved. Small calculated

values of n_T may simply correspond to small amounts of stagnant pore fluid in the rock. Large calculated values of n_T, equivalent to several rock volumes fluid, require that rocks interacted with a through-flowing fluid phase as the mineral–fluid reactions proceeded. The consequences of Eq. (2b) offer a simple criterion for identifying metamorphic or hydrothermally altered rocks that have chemically interacted with fluid that they were initially out of equilibrium with: They are rocks that record equilibrium with a fluid whose composition is different from the composition of the volatile mixture evolved by the metamorphic or hydrothermal mineral–fluid reactions. The criterion suggests that these rocks should be very common and that chemical interaction between rocks and fluids may be an almost ubiquitous phenomenon during metamorphism. Accordingly, the case studies below share a common theme: They concern metamorphosed or hydrothermally altered rocks that were in chemical equilibrium with a fluid whose composition was different from that of fluid generated within the rocks by metamorphic or hydrothermal reaction.

Regionally Metamorphosed Impure Carbonate Rocks

Beaver Brook, New Hampshire

Rumble *et al.* (1982) studied fossiliferous wollastonite-rich rocks from a single outcrop in Beaver Brook, New Hampshire. Peak pressure–temperature conditions were inferred to be 3500 bars (13 km) and 600°C. They assumed that all wollastonite was produced by the reaction:

$$\text{calcite} + \text{quartz} = \text{wollastonite} + CO_2, \qquad (11)$$

and that reaction (11) was the sole reaction that occurred during metamorphism. The progress of reaction (11) was measured by the modal abundance method. The equilibrium fluid, calculated from the calcite–quartz–wollastonite–fluid equilibrium, was H_2O rich ($X_{CO_2} = 0.09$) while fluid produced by reaction (11) was pure CO_2. During metamorphism, the carbonate rocks evidently interacted with an H_2O-rich fluid as reaction (11) progressed. Rumble *et al.* chose pure H_2O as the composition of fluid before mineral–fluid reaction and the composition of the equilibrium fluid as the composition of fluid after reaction. The amount of fluid involved in metamorphism of the wollastonite-rich rocks, calculated from Eq. (2) with $i = CO_2$ and converted to volume units, was 4.4 rock volumes.

Vassalboro Formation, South-Central Maine

Ferry (1983b) summarized the progressive metamorphism of impure carbonate rocks of the Vassalboro Formation in south-central Maine (Fig. 1). Pressure–temperature conditions during metamorphism were near 3500 bars (13 km) and between 400°C at the biotite isograd and 520–575°C in the diopside zone. Eight prograde dehydration–decarbonation reactions were identified

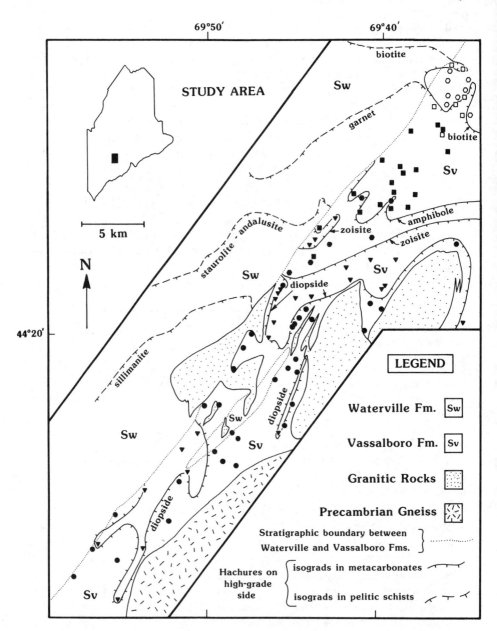

Fig. 1. Geologic sketch map of a portion of the Waterville and Vassalboro Formations, south-central Maine. Symbols are the locations of outcrops sampled: open hexagons, ankerite zone; open squares, lower biotite zone; filled squares, upper biotite zone; filled hexagons, amphibole zone; filled triangles, zoisite zone; circles, diposide zone. Metamorphic zones in the Vassalboro Formation are separated by isograds based on mineral assemblages in impure carbonate rocks (solid lines with hachures); isograds in the Waterville Formation are based on mineral assemblages in pelitic schists (dashed lines with hachures). From Ferry (1983b).

that produced biotite, anorthite, calcic amphibole, zoisite, microcline, and diopside. The reactions evolved a CO_2-rich fluid ($X_{CO_2} > 0.5$), yet various mineral equilibria record that the carbonate rocks were in chemical equilibrium with an H_2O-rich fluid ($X_{CO_2} < 0.3$) during the metamorphic event. The metacarbonate rocks must have interacted chemically with an H_2O-rich fluid as the prograde reactions proceeded. The amount of fluid involved was calculated from Eq. (3) with $i = CO_2$ by assuming that the fluid before reaction was pure H_2O, by assuming the fluid after reaction had composition of fluid with which the rocks were in equilibrium during the metamorphic event, and by measuring progress of the prograde reactions with the modal abundance and phase composition methods. Values of n_T, recalculated to volume units and expressed as a volumetric fluid–rock ratio, are summarized in Fig. 2. Taking the biotite isograd as an arbitrary zero baseline, fluid–rock ratios increase continuously with increasing metamorphic grade up to values of at least 3.1 at the highest grades. A large increase in fluid–rock ratio between ~0.2 and ~1.0 occurs abruptly at the boundary between the lower and upper biotite zones. At higher metamorphic grades, the increase in fluid–rock ratio with increasing grade is much more gradual.

Eq. (3) can be used not only to calculate fluid–rock ratios integrated over the entire metamorphic event, but also to calculate fluid–rock ratios integrated over other time intervals in the metamorphic event. At the highest grades of metamorphism in the Vassalboro Formation, the principal prograde reactions were those that produced zoisite and diopside. Applying Eq. (3) to the high-grade metacarbonate rocks from the Vassalboro Formation by summing over the effects of the diopside- and zoisite-forming reactions, fluid–rock ratios were computed just for that interval of the metamorphic event during which the two reactions proceeded. Results are summarized in Fig. 3. For individual samples at high grades, the fluid–rock ratio in Fig. 2 is much greater than the corresponding value in Fig. 3. The difference implies that most fluid that flowed through the high-grade metacarbonates did so during an early stage of the metamorphic event. Fluid continued to interact with the carbonate rocks during high-grade metamorphism, but smaller amounts of fluid were involved.

Wepawaug Formation, South-Central Connecticut

Tracy *et al.* (1983) described metamorphism of impure carbonate rocks adjacent to a synmetamorphic quartz vein in a single outcrop of the Wepawaug Formation in south-central Connecticut. Peak pressure–temperature conditions were near 6000 bars (22 km) and 520°C. Inferred prograde mineral–fluid reactions near the vein produced a CO_2-rich mixture of volatiles ($X_{CO_2} > 0.75$, their equations 9–11) while mineral assemblages record that rocks were in chemical equilibrium with an H_2O-rich fluid ($X_{CO_2} < 0.15$). Fluid-rock interaction occurred as the prograde reactions proceeded. Progress of the reactions was measured by the modal abundance and phase composition methods, and the amount of fluid with which the rocks interacted was calculated using Eq. (3) with $i = CO_2$ and the same assumptions used in the

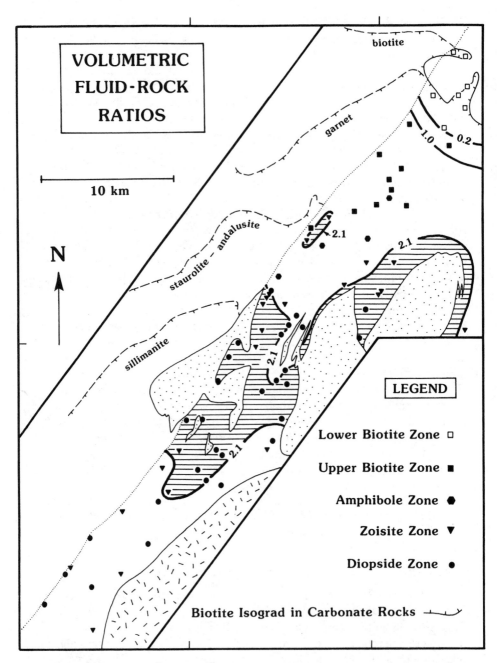

Fig. 2. Regional pattern of volumetric fluid–rock ratios during metamorphism of the Vassalboro Formation. Symbols are the data set used to determine the contours. Fluid–rock ratios integrated over the entire metamorphic event. Other features same as in Fig. 1. From Ferry (1983b).

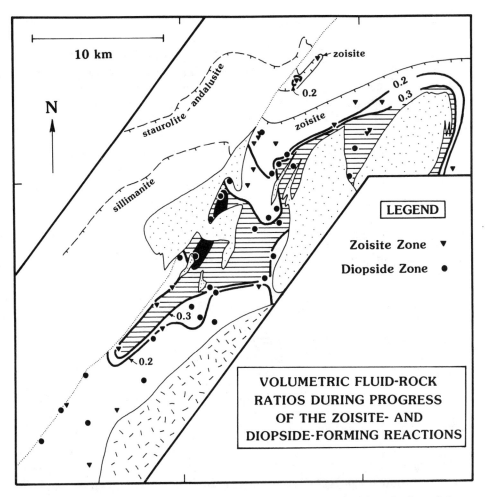

Fig. 3. Regional pattern of volumetric fluid–rock ratios during that portion of the metamorphic event during which the zoisite- and diopside-forming reactions proceeded in rocks from the Vassalboro Formation. Symbols are the data set used to determine the contours. Black area refers to a region in which fluid–rock ratios were >0.5. Area is an enlargement of part of Fig. 2. From Ferry (1983b).

studies of metacarbonate rocks at Beaver Brook and from the Vassalboro Formation. Tracy *et al.* report that some samples of Wepawaug calc-silicate rock interacted with up to at least 2 rock volumes H_2O fluid when changes in rock volume caused by the mineral–fluid reactions are adequately accounted for. The large values were not characteristic of metamorphism of the Wepawaug Formation on a regional scale, but only within several centimeters adjacent to the quartz vein, which presumably was a locus of high fluid flux during metamorphism.

Regionally Metamorphosed Pelitic Rocks

Prograde mineral reactions at a biotite isograd in pelitic phyllites from the Waterville Formation, south-central Maine (Fig. 1), produced biotite, plagioclase, and ilmenite at the expense of muscovite, ankerite, siderite, rutile, pyrite, and graphite (Ferry, 1984b). Temperature during metamorphism was near 400°C. The reactions evolved a CO_2-rich CO_2-H_2O-H_2S fluid ($X_{CO_2} >$ 0.6) yet mineral–fluid equilibria indicate that the rocks were in chemical equilibrium with fluids that were almost binary CO_2-H_2O fluids with $X_{CO_2} =$ 0.02–0.04. Pelitic rocks must have interacted with large volumes of aqueous fluid as the biotite-forming reactions proceeded. The amount of fluid involved was estimated from Eq. (3) using $i = CO_2$, assuming the fluid before reaction was pure H_2O, assuming the fluid after reaction was the equilibrium metamorphic fluid, and measuring reaction progress with the modal abundance method. At the biotite isograd all samples of pelitic phyllite chemically interacted with at least 1–2 rock volumes H_2O fluid.

Regionally Metamorphosed Graphitic Sulfidic Schists

Progressive metamorphism of graphitic, sulfide-rich schists of the Waterville Formation (Fig. 1) involved the formation of pyrrhotite at the expense of pyrite by the reactions:

$$2FeS_2 + 2H_2O + C = 2FeS + 2H_2S + CO_2, \qquad (12a)$$

and

$$2FeS_2 + CH_4 = 2FeS + 2H_2S + C, \qquad (12b)$$

at temperatures of 400–575°C (Ferry, 1981). Although the mineral–fluid reactions produced an H_2S-rich mixture of volatiles, mineral assemblages record that the schists were in equilibrium during metamorphism with fluids characterized by $X_{H_2S} = 0.02$–0.05, and fluid–rock interaction therefore accompanied reaction. The amount of fluid was estimated from Eq. (2) with $i = H_2S$ by assuming that fluid before reaction contained no H_2S ($X^0_{H_2S} = 0$), by assuming fluid composition after reaction was that of fluid in equilibrium with the rocks, and by measuring reaction progress with the modal abundance method. Results show that the graphitic sulfide-rich schists chemically interacted with at least 2–5 rock volumes CO_2-CH_4-H_2O fluid as reactions (12) proceeded during metamorphism.

Contact Metamorphosed Impure Carbonate Rocks

Nabelek *et al.* (1984) and Hover-Granath *et al.* (1983) studied progressive contact metamorphism of impure carbonate rocks of the Cambrian Orr Formation against the Jurassic Notch Peak granite. Pressure during metamorphism was near 1500 bars (6 km), and most mineral reaction occurred between approximately 500° and 600°C. In their high-grade, so-called Group 1 and Group 2 rocks, diopside, wollastonite, grossular-rich garnet, vesuvianite, and scapolite developed by a series of reactions that liberated a CO_2-rich CO_2-H_2O fluid ($X_{CO_2} > 0.5$). Mineral assemblages in the Group 1 and Group 2 rocks record equilibrium at 500–600°C with H_2O-rich fluids with $X_{CO_2} = $ ~0.05–0.3 and ~0.04–0.05, respectively. The impure carbonate rocks interacted with aqueous fluids during contact metamorphism, and the amount of fluid involved was estimated with $i = CO_2$. The composition of fluid before reaction was taken as pure H_2O ($X_{CO_2} = 0$). The value of $X^f_{CO_2}$ was taken as 0.05–0.3 and 0.04–0.05, respectively, for Group 1 and Group 2 rocks. Values of $\Sigma_k \nu_{i,k}\xi_k$, calculated from carbon isotope data of Nabelek *et al.* (their Table 3, column 9), are in the range 6.6–17.2 and 6.6–18.5 mol/L of metamorphic rock for Groups 1 and 2, respectively. For large n_T (Eq. (3)),

$$n_T + \sum_k \left[\sum_j (\nu_{j,k}\xi_k) \right] \cong n_T + \sum_k (\nu_{i,k}\xi_k).$$

Values of n_T, calculated from Eq. (3) and the above approximation and converted to rock volume units, are in the range 0.5–10.4 and 5.0–14.1 for Group 1 and Group 2 samples, respectively.

Contact Metamorphosed Pelitic Rocks

Labotka *et al.* (1984) studied progressive contact metamorphism of pelitic phyllites of the Precambrian Rove Formation against the Duluth Complex, northeastern Minnesota. Pressure during metamorphism was near 1500 bars (6 km); two mineral reactions produced cordierite, biotite, and K-feldspar in the temperature interval 500–600°C during metamorphism. The reactions liberated a pure H_2O fluid, yet mineral assemblages record equilibrium at 500–600°C with CO_2-H_2O fluids with $X_{H_2O} = 0.15$–1.00. Fluid–rock interaction occurred while the prograde reactions proceeded. The amount of fluid involved was estimated by measuring the progress of the dehydration reactions with the phase composition method, assuming the fluid after reaction had composition of CO_2-H_2O fluid with which the rocks were in equilibrium ($X^f_{H_2O} = 0.15$–1.00), assuming the fluid before reaction was pure CO_2 ($X^0_{H_2O} = 0$), and applying Eq. (3) with $i = H_2O$. Calculated volumetric fluid–rock ratios decreased from ~0.6 to 0 as temperature increased from 500° to 600°C during the contact metamorphic event. The source of the CO_2 was inferred to be metamorphic decarbonation reactions in the underlying Gunflint Iron Formation.

Hydrothermal Metamorphism of Calcareous Sandstones

Active hydrothermal metamorphism of calcareous sandstones is occurring at depth in the Salton Sea geothermal field (Muffler and White, 1969; Bird and Norton, 1981; McDowell and Elders, 1983). With increasing depth (and temperature), McDowell and Elders defined a "Porous Zone of chlorite grade," a "Hornfels Zone of biotite grade," and a "Hornfels Zone of garnet grade." At or near the boundaries of the chlorite zone, chlorite, epidote, and white mica form at the expense of calcite, dolomite/ankerite, and mixed-layer silicates. Most ankerite, dolomite, and mixed-layer silicates react to form chlorite at or near the low-temperature limit of the chlorite zone while epidote typically develops at or near the high-temperature limit of the chlorite zone (D. K. Bird, 1985, personal communication). In the biotite zone, biotite forms at the expense of chlorite, white mica, and carbonate. Mineral–fluid reactions in the Salton Sea geothermal field evidently involve decarbonation. Muffler and White (1969) presented two model reactions by which chlorite and epidote may form:

$$4 \text{ dolomite} + 5 \text{ ankerite} + 2.5 \text{ kaolinite} + Fe^{3+} + H_2O + 1.5 \ O^{-2}$$
$$= \text{chlorite} + 9 \text{ calcite} + 9 \ CO_2 \qquad (13)$$

(at or near the low-temperature limit of the chlorite zone), and

$$\text{muscovite} + 4 \text{ calcite} + 6 \text{ quartz} + 2 \ Fe^{3+} + 3 \ O^{-2}$$
$$= 2 \text{ epidote} + 2 \text{ K-feldspar} + 4 \ CO_2 + H_2O \qquad (14)$$

(at or near the high-temperature limit of the chlorite zone). The reactions produce a CO_2-rich fluid ($X_{CO_2} > 0.8$), yet the fluid that metamorphic rocks in the chlorite zone are in equilibrium with are characterized by CO_2 fugacities of 10–15 bars (Bird and Norton, 1981). At 1500 m depth in rock of density 2.7 g/cm^3, assuming ideal gas behavior, the CO_2 fugacities correspond to values of $X_{CO_2} = {\sim}0.025{-}0.038$. The difference in composition between the fluid produced by reactions (13) and (14) and the composition of fluid with which the metamorphic rocks are in equilibrium is a petrologic signature of extensive chemical interaction between rock and fluid at depth in the geothermal field. If well-developed patterns for chlorite and epidote are observed in powder X-ray diffractograms of rocks from the Salton Sea field that experienced both reactions (13) and (14) (Muffler and White, 1969), the two minerals are probably present in amounts $>{\sim}10$ wt%. After converting these weight fractions into numbers of moles of minerals per liter of metamorphic rock with density 2.7 g/cm^3, minimum estimates of the progress of reactions (13) and (14) were calculated using the modal abundance method: $\xi_{13} > 0.224$; $\xi_{14} > 0.279$ (mol/L). Minimum amounts of fluid associated with metamorphism of such rocks were then calculated from Eq. (3) assuming that only reactions (13) and (14) occur; assuming $X_{CO_2}^0 = 0$; assuming that $X_{CO_2}^f = 0.03$ (i.e., assuming reactions (13) and (14) proceeded in the presence of fluid with which rocks from the chlorite zone are in equilibrium); and utilizing the minimum estimates of ξ_{13} and ξ_{14}. The result is 2.6 rock volumes H_2O fluid.

Active hydrothermal metamorphism in the Cerro Prieto geothermal field produces calc-silicate minerals, including diopside, wollastonite, and Ca-garnet, at the expense of carbonate cement in sandstones at 300–350°C and at depths of ${\sim}1500$ m (Bird et al., 1984; Schiffman et al., 1984). Mineral-fluid reactions at Cerro Prieto evidently also involve extensive decarbonation.

The fluids with which rocks at Cerro Prieto are in equilibrium are even more impoverished in CO_2 than those in the Salton Sea geothermal field (Bird *et al.*, 1984). The stabilization of wollastonite over calcite + quartz at 325°C and 1500 m depth, for example, requires coexisting fluids with $X_{CO_2} \ll 0.01$. Progress of decarbonation reactions in the presence of nearly pure H_2O fluid at Cerro Prieto is firm petrologic evidence for active chemical interaction between rock and geothermal fluid at depth. Because of the lower values of X_{CO_2} for fluids at Cerro Prieto, fluid–rock ratios are probably somewhat higher than those estimated for the Salton Sea geothermal field. In any case, it is clear that the fluid–rock ratios associated with hydrothermal metamorphism are not significantly different from those associated with numerous instances of contact and regional metamorphism. The Salton Sea and Cerro Prieto fields may be more general models for metamorphism than is generally accepted and, as a corollary, fluid–rock interaction may play a much more important role in "normal" metamorphism than is generally appreciated.

Hydrothermally Altered Gabbro

During the initial stage of the hydrothermal alteration of olivine augite gabbros from the Isle of Skye, northwest Scotland, primary olivine was converted to talc and magnetite by hydration–oxidation reactions such as:

$$4 \ Mg_{1.5}Fe_{0.5}SiO_4 + 6.24 \ H^+ = Mg_{2.88}Fe_{0.12}Si_4O_{10}(OH)_2 + 0.627 \ Fe_3O_4$$
$$+ \ 1.493 \ H_2O + 0.627 \ H_2 + 3.12 \ Mg^{2+} \qquad (15)$$

(Ferry, 1985). The mineralogical alteration was produced by the same hydrothermal event in which heated meteoric water altered the stable isotope composition of Tertiary igneous rocks over an area of $\sim$100 km^2 on Skye (Forester and Taylor, 1977). Alteration of olivine occurred at pressure–temperature conditions near 500 bars (1–2 km) and 550°C. The reaction in some samples of gabbro was slightly different from (15) as a function of olivine composition. Hydration–oxidation reactions evolved fluid with X_{H_2} = $\sim$0.12 during the hydrothermal event. The composition of fluid in equilibrium with altered gabbro was calculated assuming that it was for practical purposes an O-H fluid and using the compositions of secondary magnetite and ilmenite in the rocks. The equilibrium fluid was characterized by X_{H_2} = 0.002–0.004. The difference between the equilibrium fluid composition and the composition of fluid evolved by the hydrothermal reaction is a petrologic signature of chemical interaction between rock and fluid during the hydrothermal event. The amount of fluid that interacted with individual samples of altered gabbro was calculated with Eq. (2) by using i = H_2, by assuming that the fluid before reaction had composition $X_{H_2}^0 = 0$ (reasonable for meteoric water in equilibrium with the atmosphere), by assuming that the fluid after reaction had composition $X_{H_2}^0 = 0.002$–0.004, and by measuring reaction progress with the modal abundance method. Values of n_T, calculated from Eq. (2) and converted into rock volume units, are in the range 0.2–6.1.

Discussion

Monitoring Fluid–Rock Interaction with Stable Isotope Data and Reaction Progress: Comparison and Significance

Comparison of Basic Equations

Closed system water–rock ratios (W/R) for fluid–rock interactions are conventionally calculated from stable isotope data utilizing the equation:

$$(W/R) = \frac{\delta^f_{rock} - \delta^0_{rock}}{\delta^0_{H_2O} - (\delta^f_{rock} - \Delta)} \qquad (16)$$

(Taylor, 1977), where W/R is in oxygen units, δ is $\delta^{18}O$, superscript 0 refers to conditions before reaction, superscript f refers to conditions after reaction, and $\Delta = \delta_{rock} - \delta_{H_2O}$. Open system water–rock ratios are $\ln(W/R + 1)$ with W/R calculated from Eq. (16). Eq. (2) may be considered the mineralogic counterpart of Eq. (16) and Eqs. (2) and (16) are closely analogous in a number of ways. First, the two equations are simply statements of mass balance. Second, in both cases, amounts of fluid are estimated based on the progress of a mineral–fluid reaction. With isotope data, a mineral–fluid isotope exchange reaction is considered; with petrologic data, a devolatilization reaction is considered. Third, both equations require the same input: (1) a measure of the progress of the mineral–fluid reaction, (2) an estimation of the composition of fluid before mineral–fluid reaction, and (3) an estimation of the composition of fluid after reaction. Reaction progress is explicitly entered into Eq. (2) as ξ; the quantity $(\delta^f_{rock} - \delta^0_{rock})$ in Eq. (16) is the implicit measure of progress of the isotope exchange reaction. The composition of fluid before reaction is X^0_i in Eq. (2) and $\delta^0_{H_2O}$ in Eq. (16). Estimates of the composition of fluid after reaction in both cases is taken as the composition of fluid with which altered rock was in chemical equilibrium. In Eq. (16) the equilibrium fluid composition is entered implicitly as $(\delta^f_{rock} - \Delta)$; in Eq. (2) X^f_i is calculated separately from mineral–fluid equilibria. The close similarity between Eqs. (2) and (16) emphasizes that fluid–rock ratios measured from isotopic data and from petrologic data have closely analogous meanings.

Comparison of Fluid–Rock Ratios from Isotopic Data with Those from Petrologic Data

In five of the case studies summarized earlier, there is sufficient stable isotope data for fluid–rock ratios to be calculated. The isotopic fluid–rock ratios provide an independent confirmation of fluid–rock ratios calculated from petrologic data and reaction progress. Rumble et al. (1982) calculated open system isotopic fluid–rock ratios for the wollastonite-bearing rocks at Beaver Brook, New Hampshire, by assuming that rocks before metamorphism had the isotopic composition of typical silicified Devonian brachiopods and that fluid before reaction was H_2O in isotopic exchange equilibrium with the pelitic schists that surround the calc-silicate rocks. Results are 5–6 rock volumes H_2O, compared to 4–5 rock volumes estimated from petrologic data. The rocks of the Wepawaug Formation, south-central Connecticut, have oxygen isotope composition so thoroughly homogenized by metamorphic fluid–rock interaction that isotopic data only provide a lower bound on the amount of fluid involved (Tracy et al., 1983). The lower bound is approximately 4–5 rock volumes H_2O, compared to ~2 rock volumes estimated from petrologic data. Nabelek et al. (1984) estimated open-system isotopic fluid–rock ratios for calc-silicate rocks developed during contact metamorphism by the Notch Peak granite, Utah. The isotopic composition of fluid before reaction was taken as that of H_2O in isotopic exchange equilibrium with the granite, and the composition of the rock before reaction was taken as unmetamorphosed calcareous argillite in the area corrected for the effect of Rayleigh distillation during the mineral–fluid reaction. Results are approximately 1–5 and 2–4 rock volumes, respectively, for Group 1 and Group 2 samples compared to volumetric fluid–rock ratios of 1–10 and 5–14 for the two groups, respectively, estimated from petrologic data. Clayton et al. (1968) calculated a closed system isotopic fluid–rock ratio for rocks from the Salton Sea geothermal field assuming the isotopic composition of fluid before water–rock interaction was that of local meteoric water and that rock before reaction had composition of unaltered sediments in the area; results are 1.2 rock volumes H_2O compared to values of >~2.6 calculated from petrologic data. Closed-system water–rock ratios were calculated for the hydrothermal event that altered gabbros from the Isle of Skye, Scotland, utilizing the equation in Table 2 of Forester and Taylor (1977). The isotopic composition of fluid before reaction was taken as their inferred value for local Tertiary meteoric water, and the composition of rock before reaction was taken as their inferred value for unaltered gabbro. Results are 1–10 rock volumes H_2O, compared to volumetric fluid–rock ratios of 0.2–6 calculated from petrologic data. Although the agreement between fluid–rock ratios calculated from isotopic data and from petrologic data is only within approximately a factor of two, the agreement is encouraging considering the different sorts of data used and the number of assumptions and approximations used in each technique. The most important conclusion to be made by comparing the results of the two methods is that both record chemical interaction of rock with several rock volumes fluid during hydrothermal and regional and contact metamorphic events.

Significance of Comparison between Petrologic and Isotopic Fluid–Rock Ratios

The difference between the composition of fluid with which the rocks from the various case studies were in equilibrium and the composition of fluid that the rocks evolved by mineral–fluid reactions has been interpreted in terms of the chemical interaction of rock with fluids as the reactions proceeded. The difference in fluid compositions could also, in principle, be explained by differential loss of volatile components from the rock during reaction with no chemical interaction of rock with externally derived fluids. For example, during metamorphism of carbonate rocks, the difference between the CO_2-rich composition of fluid generated by the prograde mineral reactions and the H_2O-rich composition of fluids with which the minerals record equilibrium could be explained if CO_2 molecules produced by the reactions escaped more readily than did the H_2O molecules. The H_2O molecules could form an H_2O-rich residue with which the minerals in the rocks would then equilibrate. The integrated isotopic and petrologic studies of Rumble *et al.* (1982), Tracy *et al.* (1983), and Nabelek *et al.* (1984) permit a process of fluid–rock interaction to be unequivocally distinguished from a process of preferential CO_2 loss during metamorphism. Preferential CO_2 loss can change a rock's isotopic composition only by Rayleigh distillation, and Rayleigh distillation generally leaves an isotopic signature greatly different from that left when extensive fluid–rock interaction accompanies mineral–fluid reaction. Specifically, Rayleigh distillation can alter the oxygen isotope composition of a rock undergoing a normal decarbonation–dehydration reaction by no more than ~1–2 per mil $\delta^{18}O$ (Rumble 1982; Rumble *et al.*, 1982). Rocks studied by Rumble *et al.* (1982), Tracy *et al.* (1983), and Nabelek *et al.* (1984) changed in $\delta^{18}O$ by 3–15 per mil, and these large changes in whole-rock oxygen isotope composition are an irrefutable demonstration that fluid–rock interaction rather than differential loss of CO_2 occurred during the metamorphic event. In every case where the interpretation of petrologic data in terms of fluid–rock interaction has been critically tested by isotopic studies, the model for fluid–rock interaction has been supported. In no instance have isotopic data demonstrated that preferential loss of CO_2 (or any other component) from the fluid occurred during metamorphism.

Fluid Flow in Rocks at Depths in the Crust Greater than 10 Kilometers

Evidence for Fluid Flow at Depths >10 Km

Case studies by Rumble *et al.* (1982), Tracy *et al.* (1983), and Ferry (1981, 1983b, 1984b) present firm isotopic and petrologic evidence for the chemical interaction of rock with at least 1–5 rock volumes fluid at depths of 13–22 km in the crust during metamorphism. The large calculated fluid–rock ratios are difficult to reconcile with transport of volatile components by diffusion in a stagnant pore fluid. Large calculated fluid–rock ratios are better interpreted in terms of flow of fluid through rocks. Values of fluid–rock ratios for some rock samples from deep-seated metamorphic terranes are similar in magnitude to those associated with hydrothermal convection cells driven by epizonal plutons (Taylor, 1977; Norton and Taylor, 1979). The similarity in values suggests that mesozonal and catazonal hydrothermal convection cells may also exist during regional metamorphism. Regional patterns of high fluid–rock ratios concentrically centered on deep-seated synmetamorphic plutonic intrusions (e.g., Figs. 2–3) further argue for a consideration of metamorphic hydrothermal cells at midlevels of the crust.

Mechanisms of Fluid Flow at Depths >10 Km

Two mechanisms of fluid flow are likely to occur in deep-seated rocks: flow along fractures and pervasive flow along grain boundaries (Walther and Wood, 1984). Walther and Orville (1982) have forcefully argued that fluid flow occurs during metamorphism along vertical or near vertical fractures. While not discounting fluid flow along fractures, petrologic observations, visible in outcrop, argue for a significant component of metamorphic fluid flow that is pervasive on a grain-size scale. Because reaction progress is a measure of the amount of fluid that rock has chemically interacted with during metamorphism, the distribution of rocks in outcrop that have undergone mineral–fluid reaction is a visual record of where fluid has flowed. In pelitic schists from the biotite zone of the Waterville Formation, in impure carbonate rocks of the Vassalboro Formation, and in calc-silicate rocks from Beaver Brook, mineral–fluid reactions have pervasively affected all rock within beds of a particular lithologic type. The uniformity of reaction progress seen within beds in outcrop has been verified quantitatively by determination of proportions of prograde minerals in thin sections prepared from different portions of individual beds. The distribution of reaction progress in outcrop is evidence for pervasive fluid–rock interaction on a grain size scale. Reaction progress, however, may vary between beds of the same outcrop. At least some pervasive metamorphic fluid flow therefore must be channelized within beds, analogous to the flow of groundwater through rocks close to the earth's surface. The evidence for pervasive flow, channelized along beds during regional metamorphism, is in harmony with the study of Nabelek et al. (1984). Their isotopic data unequivocally demonstrate that fluid flow in the contact aureole of the Notch Peak granite was pervasive but channelized within horizontal beds of calcareous argillite. The argillites are separated by massive beds of relatively pure carbonate rocks whose isotopic composition requires that they were relatively impermeable during the metamorphic event. The channelization of fluid flow along bedding during metamorphism can be understood in terms of a model for reaction-enhanced permeability (Rumble et al., 1982). In a heterogeneous sequence of sediments, some beds will have bulk chemical and mineralogical composition favorable for extensive progress of metamorphic devolatilization reactions (the calcareous argillites in the contact aureole of the Notch Peak granite, for example). Because devolatilization reactions (especially decarbonation reactions) enhance permeability and hence fluid flow by increasing porosity and maintaining high values of fluid pressure, beds that undergo significant reaction during metamorphism will become loci of high fluid flow—metamorphic aquifers. In contrast, beds with bulk chemical and mineralogical composition unfavorable to extensive reaction (the massive pure carbonate rocks in the contact aureole of the Notch Peak granite, for example) will not experience reaction-enhanced permeability during metamorphism and will develop into relatively impermeable metamorphic aquitards.

The chlorite zone is a region of enhanced fluid flow in the Salton Sea geothermal field. McDowell and Elders (1983) report a sudden increase in porosity of ~5–20% with depth at the beginning of the chlorite zone. The increase in porosity can be explained by the change in rock volume caused by active devolatilization reactions that convert mixed-layer silicates and carbonates to chlorite in the chlorite zone. The increase in porosity in the chlorite zone, in turn, explains the greater fluid flow at that depth, i.e., active reaction enhancement of permeability is probably occurring in the Salton Sea geothermal field.

Significance of Pervasive Fluid Flow through Rocks during Regional Metamorphism

The pervasive flow of fluid through rocks during metamorphism has profound consequences for the understanding of the process of metamorphism. When fluid flows through rock and they are not in chemical equilibrium, reactions occur between the fluid and minerals in the rock as they attempt to attain chemical equilibrium with each other. Fluid flow through rocks therefore causes metamorphic mineral reactions to occur and acts as a fundamental driving force behind metamorphism. The measurement of fluid–rock interaction by reaction progress with Eqs. (2) and (3) is essentially a measurement of the degree to which mineral reactions in a metamorphic rock were driven by chemical interaction between rock and fluid during the metamorphic event. Conventionally, absorption of heat by rocks is considered the only significant driving force behind metamorphism. The extensive amount of metamorphic fluid flow required by the case studies summarized earlier is persuasive testimony that fluid–rock interaction, in addition to heat–rock interaction, should always be considered before the petrogenesis of a suite of metamorphic rocks can be adequately understood. The sharp increase in fluid–rock ratio at or near the biotite isograd in both the Waterville and Vassalboro Formations (Fig. 1) suggests that fluid flow and consequent fluid–rock interaction may even act as the harbinger of some regional metamorphic events. Even in instances of hydrothermal metamorphism, such as in the Salton Sea and Cerro Prieto geothermal fields, metamorphism is sometimes attributed primarily to anomalously high geothermal gradients. The large fluid–rock ratios inferred for metamorphism in these two areas, however, is firm evidence that chemical interaction between rock and fluid is an essential driving force behind the active metamorphism. The mineralogical changes observed with depth in the Salton Sea and Cerro Prieto fields results from the departure of rocks from both chemical and thermal equilibrium with fluid that flows through them.

As this review indicates, most of the evidence for pervasive fluid–rock interaction during metamorphism is derived from the study of impure carbonate rocks. Metacarbonates have been extensively studied because the most common fluid at depth in metamorphic terranes evidently is H_2O-rich, and the interaction of impure carbonate rocks at high temperature with aqueous fluid readily drives decarbonation reactions in them. Fluid–rock interaction leaves a dramatic mineralogical record in metacarbonate rocks. Other common lithologies, normal pelitic schists and metamorphosed volcanic rocks, for example, have not been extensively studied for petrologic evidence of fluid–rock interaction during metamorphism. The reason is that pelitic schists and amphibolites principally undergo dehydration reactions during prograde metamorphism, and dehydration reactions cannot be driven by interaction of rock with H_2O fluid. Consequently, fluid–rock interaction generally leaves little or no petrologic signature on these rock types. An important problem remains to determine whether pervasive fluid flow occurs in all rock types in metamorphic terranes or is typically restricted to impure carbonate beds. The problem is likely to be resolved more effectively with stable isotope rather than with petrologic data.

The alkali content of impure carbonate rocks from the Vassalboro Formation, south-central Maine, is plotted in Fig. 4 against their calculated volumetric fluid–rock ratios. The negative correlation implies that depletion of the rocks in both K and Na during metamorphism was effected by large volumes of through-flowing metamorphic fluid. Tracy *et al.* (1983) also document a negative correlation between fluid–rock ratio and alkali content of calc-silicate rocks from the Wepawaug Formation. Evidently fluid flow through rocks during metamorphism can have a profound effect on rock chemistry and hence on mineralogy. Not all instances of high metamorphic fluid flux, however, are associated with allochemical metamorphism. Metamorphism of pelitic schists at the biotite isograd of the Waterville Formation and of the calcareous argillites in the contact aureole of the Notch Peak granite, while involving large fluid–rock ratios, was for practical purposes isochemical (Ferry, 1984b; Hover-Granath *et al.*, 1983). An important problem is to understand why some instances of metamorphic fluid flow are associated with mobilization and transfer of significant amounts of major rock-forming elements and why other instances of fluid flow result in extensive devolatilization at constant or nearconstant rock composition.

The heat budget of impure carbonate rocks from the Vassalboro Formation is plotted in Fig. 5 against their calculated volumetric fluid–rock ratios. The positive correlation implies that a significant amount of heat was transferred during metamorphism in the area by a through-flowing fluid. Because fluid flow in rocks normally occurs at a faster rate than heat flow by conduction (e.g., Norton and Knight, 1977), metamorphic fluid flow must play a part in metamorphic heat transfer and must, at least in part, control the thermal structure and thermal evolution of metamorphic belts. Conventional treatments of the energy budget of metamorphism assume that heat transfer occurs exclusively by conduction (e.g., England and Richardson 1977; En-

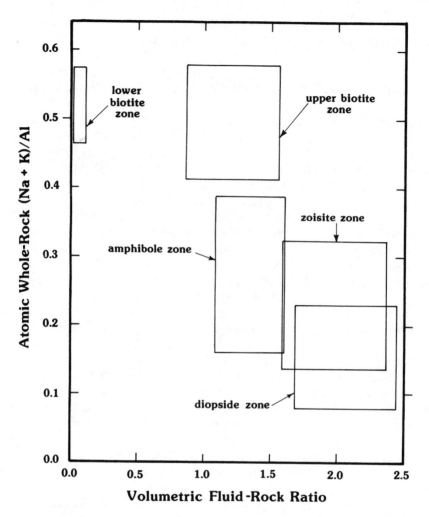

Fig. 4. Relationship between calculated volumetric fluid–rock ratio and the measured atomic whole-rock ratio (Na + K)/Al for metamorphosed carbonate rocks from the Vassalboro Formation. Center of boxes represents the mean of plotted variables for all samples from each zone (cf. Fig. 1); length of sides of boxes represents two standard deviations about that mean. From Ferry (1983b).

gland and Thompson, 1984; Thompson and England, 1984). Such assumptions should be critically reevaluated in the light of petrologic and isotopic evidence for significant amounts of pervasive fluid flow that evidently occurs during regional metamorphism. Certainly heat flow occurs neither solely by conduction nor solely by convection. The essential question is the relative importance of the two mechanisms in the various environments of regional metamorphism. Metamorphic fluid flow may play a profound role in at least some instances of metamorphism not only as a fundamental causative agent

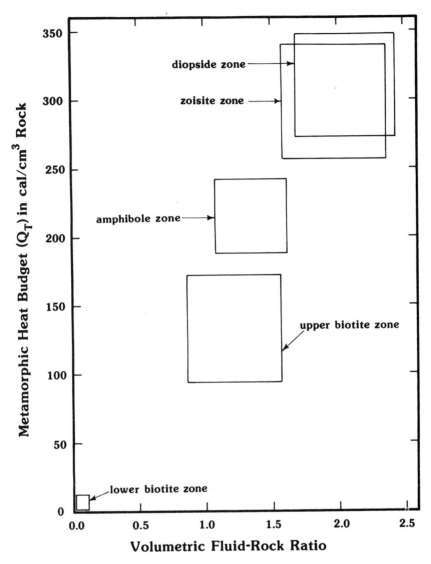

Fig. 5. Relationship between calculated volumetric fluid–rock ratio and the total heat budget (relative to conditions at the biotite isograd) for metamorposed carbonate rocks from the Vassalboro Formation. Boxes have same significance as in Fig. 4. From Ferry (1983b).

and as a medium for major element mass transfer but also as a significant controlling factor of heat transfer. As in active geothermal fields, regional metamorphism may in some or many cases be the response of rocks to a departure from both chemical and thermal equilibrium with fluid that flows through them.

Acknowledgments

Research supported by grant EAR 82-18464 from the Division of Earth Sciences, National Science Foundation. Thoughtful reviews by D. K. Bird, G. C. Flowers, and R. J. Tracy pointed out a number of misconceptions in an early version of the manuscript and were much appreciated.

References

Bird, D.K., and Norton, D.L. (1981) Theoretical prediction of phase relations among aqueous solutions and minerals: Salton Sea geothermal system. *Geochim. Cosmochim. Acta* **45**, 1479–1493.

Bird, D.K., Schiffman, P., and Elders, W.A., Williams, A.E., and McDowell, S.D. (1984) Calc-silicate mineralization in active geothermal systems. *Econ. Geol.* **79**, 671–695.

Brimhall, G.H., Jr. (1979) Lithologic determination of mass transfer mechanisms of multiple-stage porphyry copper mineralization at Butte, Montana: Vein formation by hypogene leaching and enrichment of potassium-silicate protore. *Econ. Geol.* **74**, 556–589.

Clayton, R.N., Muffler, L.J.P., and White, D.E. (1968) Oxygen isotope study of calcite and silicates of the River Ranch No. 1 well, Salton Sea geothermal field, California. *Am. J. Sci.* **266**, 968–979.

DeDonder, Th. (1920) *Leçons de Thermodynamique et de Chimie-Physique*. Gauthier-Villars, Paris.

England, P.C., and Richardson, S.W. (1977) The influence of erosion upon the mineral facies of rocks from different metamorphic environments. *J. Geol. Soc. London* **134**, 201–213.

England, P.C., and Thompson, A.B. (1984) Pressure-temperature-time paths of regional metamorphism. Part I. Heat transfer during evolution of regions of thickened continental crust. *J. Petrol.* **25**, 894–928.

Ferry, J.M. (1980) A case study of the amount and distribution of heat and fluid during metamorphism. *Contrib. Mineral. Petrol.* **71**, 373–385.

Ferry, J.M. (1981) Petrology of graphitic sulfide-rich schists from south-central Maine: An example of desulfidation during prograde regional metamorphism. *Amer. Mineral.* **66**, 908–930.

Ferry, J.M. (1983a) Applications of the reaction progress variable in metamorphic petrology. *J. Petrol.* **24**, 343–376.

Ferry, J.M. (1983b) Regional metamorphism of the Vassalboro Formation, south-central Maine, USA: A case study of the role of fluid in metamorphic petrogenesis. *J. Geol. Soc. London* **140**, 551–576.

Ferry, J.M. (1983c) On the control of temperature, fluid composition, and reaction progress during metamorphism. *Amer. J. Sci.* **283A**, 201–232.

Ferry, J.M. (1984a) Phase composition as a measure of reaction progress and an experimental model for the high-temperature metamorphism of mafic igneous rocks. *Amer. Mineral.* **69**, 677–691.

Ferry, J.M. (1984b) A biotite isograd in south-central Maine, U.S.A.: Mineral reactions, fluid transfer, and heat transfer. *J. Petrol.* **25**, 871–893.

Ferry, J.M. (1985) Hydrothermal alteration of Tertiary igneous rocks from the Isle of Syke, northwest Scotland. I. Gabbros. *Contrib. Mineral. Petrol.* **91**, 264–282.

Ferry, J.M., and Burt, D.M. (1982) Characterization of metamorphic fluid composition through mineral equilibria, in *Characterization of Metamorphism through Mineral Equilibria,* edited by J.M. Ferry, pp. 207–262. Mineral. Soc. Amer., Washington, DC.

Fitts, D.D. (1962) *Nonequilibrium Thermodynamics.* McGraw-Hill, New York.

Forester, R.W., and Taylor, H.P. (1977) $^{18}O/^{16}O$, D/H, and $^{13}C/^{12}C$ studies of the Tertiary igneous complex of Skye, Scotland. *Amer. J. Sci.* **277**, 136–177.

Helgeson, H.C. (1968) Evaluation of irreversible reactions in geochemical processes involving minerals and aqueous solutions. I. Thermodynamic relations. *Geochim. Cosmochim. Acta* **32**, 853–877.

Helgeson, H.C. (1970) A chemical and thermodynamic model of ore deposition in hydrothermal systems. *Mineral. Soc. Amer. Spec. Pap.* **3**, 155–186.

Helgeson, H.C., Garrels, R.M., and Mackenzie, F.T. (1969) Evaluation of irreversible reactions in geochemical processes involving minerals and aqueous solutions. II. Applications. *Geochim. Cosmochim. Acta* **33**, 455–481.

Helgeson, H.C., Brown, T.H., Nigrini, A., and Jones, T.A. (1970) Calculation of mass transfer in geochemical processes involving aqueous solutions. *Geochim. Cosmochim. Acta* **34**, 569–592.

Hover-Granath, V.C., Papike, J.J., and Labotka, T.C. (1983) The Notch Peak contact metamorphic aureole, Utah: Petrology of the Big Horse limestone member of the Orr Formation. *Geol. Soc. Amer. Bull.* **94**, 889–906.

Labotka, T.C., White, C.E., and Papike, J.J. (1984) The evolution of water in the contact-metamorphic aureole of the Duluth Complex, northeastern Minnesota. *Geol. Soc. Amer. Bull.* **95**, 788–804.

McDowell, S.D., and Elders, W.A. (1983) Allogenic layer silicate minerals in borehole Elmore #1, Salton Sea geothermal field, California. *Amer. Mineral.* **68**, 1146–1159.

Muffler, L.J.P., and White, D.E. (1969) Active metamorphism of Upper Cretaceous sediments in the Salton Sea geothermal field and the Salton Trough, southeastern California. *Geol. Soc. Amer. Bull.* **80**, 157–182.

Nabelek, P.I., Labotka, T.C., O'Neil, J.R., and Papike, J.J. (1984) Contrasting fluid/rock interaction between the Notch Peak granitic intrusion and argillites and limestones in western Utah: Evidence from stable isotopes and phase assemblages. *Contrib. Mineral. Petrol.* **86**, 25–34.

Norton, D.L., and Knight, J. (1977) Transport phenomena in hydrothermal systems: Cooling plutons. *Amer. J. Sci.* **277**, 937–981.

Norton, D.L., and Taylor, H.P., Jr. (1979) Quantitative simulation of the hydrothermal systems of crystallizing magmas on the basis of transport theory and oxygen isotope data: An analysis of the Skaergaard intrusion. *J. Petrol.* **20**, 421–486.

Prigogine, I., and Defay, R. (1954) *Chemical Thermodynamics.* Longman, London.

Rice, J.M., and Ferry, J.M. (1982) Buffering, infiltration, and the control of intensive variables during metamorphism, in *Characterization of Metamorphism through Mineral Equilibria,* edited by J.M. Ferry, pp. 263–326. Mineral. Soc. Amer., Washington, DC.

Rumble, D. (1982) Stable isotope fractionation during metamorphic devolatilization reactions, in *Characterization of Metamorphism through Mineral Equilibria,* edited by J.M. Ferry, pp. 327–354. Mineral. Soc. Amer., Washington, DC.

Rumble, D., Ferry, J.M., Hoering, T.C., and Boucot, A.J. (1982) Fluid flow during metamorphism at the Beaver Brook fossil locality, New Hampshire. *Amer. J. Sci.* **282,** 886–919.

Schiffman, P., Elders, W.A., Williams, A.E., McDowell, S.D., and Bird, D.K. (1984) Active metasomatism in the Cerro Prieto geothermal system, Baja California, Mexico: A telescoped low-pressure, low-temperature metamorphic facies series. *Geology* **12,** 12–15.

Spear, F.S. (1984) Contrasting P–T paths of Barrovian and Buchan metamorphism, central New England. *Trans. Amer. Geophys. Union* **65,** 1148.

Spear, F.S., and Selverstone, J. (1983) Quantitative P–T paths from zoned minerals: Theory and tectonic applications. *Contrib. Mineral. Petrol.* **83,** 348–357.

Taylor, H.P., Jr. (1977) Water/rock interactions and the origin of H_2O in granitic batholiths. *J. Geol. Soc. London* **133,** 509–558.

Thompson, A.B., and England, P.C. (1984) Pressure-temperature-time paths of regional metamorphism. Part II. Their inference and interpretation using mineral assemblages in metamorphic rocks. *J. Petrol.* **25,** 929–955.

Thompson, J.B., Jr. (1982a) Composition space: An algebraic and geometric approach, in *Characterization of Metamorphism through Mineral Equilibria,* edited by J.M. Ferry, pp. 1–32. Mineral. Soc. Amer., Washington, DC.

Thompson, J.B., Jr. (1982b) Reaction space: An algebraic and geometric approach, in *Characterization of Metamorphism through Mineral Equilibria,* edited by J.M. Ferry, pp. 33–52. Mineral. Soc. Amer., Washington, DC.

Thompson, J.B., Jr., Laird, J., and Thompson, A.B. (1982) Reactions in amphibolite, greenschist and blueschist. *J. Petrol.* **23,** 1–27.

Tracy, R.J., Rye, D.M., Hewitt, D.A., and Schiffries, C.M. (1983) Petrologic and stable-isotopic studies of fluid-rock interactions, south-central Connecticut. I. The role of infiltration in producing reaction assemblages in impure marbles. *Amer. J. Sci.* **283A,** 589–616.

Turner, F.J. (1981) *Metamorphic Petrology. Mineralogical, Field, and Tectonic Aspects.* McGraw-Hill, New York.

Walther, J.V., and Orville, P.M. (1982) Volatile production and transport in regional metamorphism. *Contrib. Mineral. Petrol.* **79,** 252–257.

Walther, J.V., and Wood, B.J. (1984) Rate and mechanism in prograde metamorphism. *Contrib. Mineral. Petrol.* **88,** 246–259.

Chapter 4
Fluid Flow during Metamorphism and Its Implications for Fluid–Rock Ratios

B.J. Wood and J.V. Walther

Introduction

The study of metamorphic rocks is traditionally approached as an investigation of mineralogical changes occurring as responses to variations only in pressure and temperature. Despite the losses of large volumes of fluid during prograde metamorphism, rocks are often regarded, on the thin section scale, as systems able to buffer the chemical potentials (μ) of all components and therefore as essentially closed. There are several important reasons, apart from its simplicity, for adherence to this model. First, there is the observation that metasediments maintain their original sharp compositional discontinuities even in the amphibolite facies (e.g., Chinner, 1960). Second, low variance assemblages (implying internally buffered μ's) are common. Finally, metasomatic changes and externally controlled activities generally lead to simple mono- or bimineralic zones such as those found in skarns or veins (e.g., Tilley, 1951; Burnham, 1959; Thompson, 1959). Such zones occur between interbedded metapelites and carbonates but are not a generally observed phenomenon in mixed pelitic/psammitic units.

The foregoing generalizations, which apply to much of the petrologic literature, envisage the fluid phase as a passive by-product of metamorphism, generated under buffered condition by reactions that produce the greatest mineralogic changes at isobaric invariant points (Greenwood, 1975). Recent detailed petrographic and isotopic studies of metamorphosed carbonate rocks (e.g., Ferry, 1983, and this volume; Rumble *et al.*, 1982; Graham *et al.*, 1983) have, however, revealed that the fluid phase plays a far more active role in prograde metamorphism of some lithologies than was previously envisaged. All of these authors have demonstrated, from both phase equilibrium and isotopic data, that greenschist and amphibolite facies metacarbonates were infiltrated by between 1 and 5 rock volumes of H_2O. These are the volumes of H_2O that have reacted with the rock and they therefore represent lower bounds to the amounts of water that actually passed through these reacting systems. Since prograde reactions in carbonate systems are dependent on H_2O/CO_2 ratio in the fluid phase and the carbonate rocks themselves generate only CO_2, it is quite possible that the observed reactions occur because of H_2O infiltration at constant temperature rather than as a response to increasing temperature at constant composition. Thus, the fluid can be an active rather than a passive agent of prograde change, stimulating reaction and enhancing rates of equilibration through its transport properties (Walther and Wood, 1984).

To date, high fluid–rock ratios have only been well documented in metacarbonate units, and it is extremely difficult to establish their implications for the overall metamorphic process. If, on the other hand, it was clearly demonstrated that large volumes of fluid have interacted with the more common pelitic and psammitic units, then the sources of this fluid and the physical processes involved in its transport would require detailed study. Unfortunately, water is the dominant fluid species evolved during prograde metamorphism and pelites produce a water-rich fluid. Therefore, the effects of externally derived fluid are difficult to document, although Ferry (1984) has some evidence of H_2O addition at the biotite isograd in the Waterville-Vassalboro area, southeast Maine. One of the aims of this paper is to consider those equilibria and mineral phases in metapelites that are sensitive to fluid–rock interactions. This provides the basis for establishing the extent of such interactions in common metasedimentary rocks. In anticipation of high fluid–rock ratios throughout metamorphic terranes, Etheridge et al. (1983) have proposed that fluid derived from metamorphic reactions convects through the metamorphosing pile, circulating and interacting many times over with each rock unit. Although such behavior has been established within and around shallow intrusions (Forester and Taylor, 1976), its operation at deep crustal levels is difficult to prove. Indeed, Magaritz and Taylor (1976) have demonstrated that there was no circulation to deep levels in the Coast Range Batholith of British Columbia. An important aim of this paper is to consider the possibility of fluid convection during regional metamorphism and to demonstrate the circumstances under which it can occur.

Fluid Production and Transport

Progressive metamorphism produces, through dehydration and decarbonation reactions, about 2 mol of fluid per kilogram of pelite (Walther and Orville, 1982), equivalent to about 130 cc/L at high pressure and temperature. Physically, this fluid must flow upward because of its low density, but, if there is an interconnected pore or fracture system, it may cool at the earth's surface and return to depth (Fig. 1(a)), continuing to circulate until the temperature gradient declines to the point at which the deep fluids are no longer buoyant. Since rocks are nearly three times as dense as H_2O-CO_2

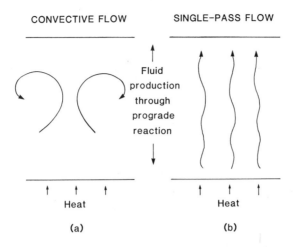

Fig. 1. Two scenarios for fluid flow attending metamorphism. (a) The rocks are strong enough to hold an interconnected pore and fracture system open and fluid may circulate. (b) The pores collapse so that fluid pressure equals lithostatic pressure. Flow can only take place toward the earth's surface.

fluids, the only circumstance under which fluid can convect is the one in which the rock is strong enough to allow a hydrostatic pressure gradient to be maintained through a system of interconnected pores and fluid pathways. The rock must therefore support a difference between the lower hydrostatic pressure of fluid within its pores and the higher lithostatic pressure exerted on it by the overlying column of rock. If the rock is of insufficient strength to hold open these pores then it will collapse until the fluid becomes pressurized to lithostatic pressure. As lithostatic pressure is approached by the fluid, convection must cease and flow will only take place toward the surface as depicted in Fig. 1(b).

The hydrostatic and lithostatic pressure gradients are given by, respectively:

$$\left|\frac{dP_h}{dz}\right| = \rho_{H_2O}g = 100 \text{ bars/km} \tag{1a}$$

$$\left|\frac{dP_l}{dz}\right| = \rho_{rock}g = 280 \text{ bars/km} \tag{1b}$$

Therefore, for every 1 km increase in depth the rock must increase in crushing strength by about 180 bars in order to maintain the hydrostatic gradient necessary for convecting fluid.

The strengths of rocks under conditions of fluid pressure and strain rate appropriate to regional metamorphism are not well known, but values on the order of tens of bars (equivalent to a few hundred meters depth) are generally thought to be appropriate at 500°C (Fyfe *et al.*, 1978, p. 223; Etheridge *et al.*, 1984). It appears unlikely, therefore, that metamorphic rocks can hold

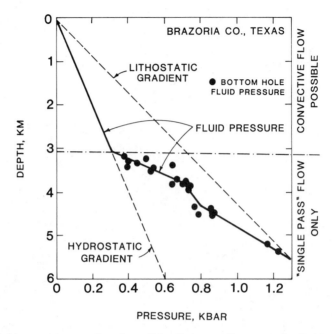

Fig. 2. Fluid pressure as a function of depth in a sedimentary basin (Gregory and Backus, 1980). Note the cross-over from hydrostatic to lithostatic fluid pressure and its implications for convection.

open a fracture or pore system over vertical distances of more than a few hundred meters at high temperatures. At lower temperatures, strengths of quartz-rich rocks increase such that at 200°C differential stresses on the order of 300 bars are needed to produce geologic strain rates (Fyfe *et al.*, 1978). This implies that, in the upper crust, the zone of possible convection might reach down a few kilometers rather than a few hundred meters.

Fig. 2 shows the measured fluid pressure with depth in near-surface sedimentary rocks from an area within the U.S. Gulf coast. As can be seen, fluid pressure follows the hydrostatic pressure gradient until a depth of 3 km is reached. At greater depths, due to collapse of the interconnected pore system, the fluid pressure gradient increases until fluid pressure approaches lithostatic pressure at 5.5 km depth. In such a case, convective flow is only possible in the depth interval over which the fluid pressure gradient is hydrostatic, i.e., 0–3 km. Rocks in this depth interval may exhibit large fluid–rock ratios particularly when convection is stimulated by an igneous intrusion (Norton and Knight, 1977). Below 3 km the fluid cannot convect, and flow can only be of the "single-pass" type (Fig. 1(b)) toward the earth's surface.

The depth at which the transition from hydrostatic to lithostatic pressure takes place varies considerably from one sedimentary basin to another, but in general it is less than 6 km (Rubey and Hubbert, 1959). Since high-grade metasedimentary rocks have all gone through the lithification and diagenesis stage en route to metamorphism, the depths of changeover from hydrostatic to lithostatic fluid pressure gradient during the prograde events should have been similar to those observed in modern sedimentary basins. Bearing in mind the observation that regional metamorphism almost invariably takes place at pressures greater than 1700 bars (6 km), it is difficult to escape the conclusion that fluid flow is usually of the single-pass type. Engelder's (1984) observations of calcite pressure solution in foreland fold and thrust belts support our conclusions. He found substantial volume losses involving circulation of meteoric water down to 1 km depth, but no evidence of circulation to greater depths.

Given that convection cannot occur and that the fluid released during metamorphism passes only once through the overlying sequence of rocks, it is of interest to investigate the effects of this "external" fluid on the rocks with which it interacts on its way.

Single-Pass Flow and Fluid–Rock Ratio in Carbonates

Consider a sequence of sedimentary rocks in which pelitic sediments dominate over limestones. During progressive metamorphism, the average pelite releases 2 mol of water-rich fluid per kilogram. A 1 km thick layer of pelite will therefore produce a total of about 530 mol of H_2O per square centimeter of surface area. In a limestone containing about 30% quartz, it is common that approximately half of the total volume of carbonate (principally dolomite) reacts to form calcium-magnesium silicates during progressive metamorphism (e.g., Flowers and Helgeson, 1983). This results in a loss of about 4 mol of CO_2 per kilogram during metamorphism.

If the H_2O generated in the pelitic part of the sequence were to flow through the reacting carbonate, it would obviously have a profound effect on the fluid composition in the latter. Turning this question around, it is appropriate to estimate what thickness of limestone would be needed in order for the apparent fluid–rock ratio to be equal to 1.0 by volume. Note that only the total integrated amount of fluid can be estimated and that this procedure requires the assumption of equilibrium between the total fluid volume and the rock.

The volume of 530 mol of H_2O per square centimeter represents a column about 110 m high. Therefore, an H_2O : carbonate ratio of 1.0 by volume results in 110 m of limestone per kilometer of pelite, or about 10% limestone in the sequence. The mole fraction of CO_2 in the resultant fluid would be about 0.17 (Fig. 3). Fig. 3 shows the approximate relationship between fluid–rock volume ratio, fluid composition, and proportion of limestone in the

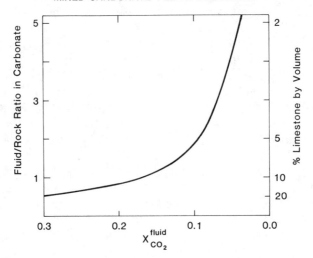

MIXED CARBONATE–PELITE SEQUENCES

Fig. 3. The influence of volumetric proportions of limestone in a sedimentary sequence on maximum observable values of $X_{CO_2}^{fluid}$ and fluid–rock ratio.

sequence. Although the amount of carbonate in metamorphic terranes is highly variable, values on the order of 5–10% with corresponding fluid–rock ratio of 1.0–2.0 and X_{CO_2} of 0.1–0.17 seem to be in reasonable agreement with the available petrologic data (e.g., Rumble *et al.*, 1982).

Apart from introducing H_2O into reacting carbonates, the dehydrating sequence of pelites can add some of the $SiO_{2(aq)}$ that is consumed by decarbonation reactions of the type:

$$5CaMg(CO_3)_2 + H_2O + 8SiO_2$$
dolomite

$$= Ca_2Mg_5Si_8O_{22}(OH)_2 + 3CaCO_3 + 7CO_2. \quad (2)$$
$$\text{tremolite} \qquad \text{calcite}$$

In general most of this will be added near the boundaries between carbonate and silicate-rich horizons since the carbonate consumes SiO_2 and the addition of CO_2 dramatically reduces SiO_2 solubility in the H_2O-CO_2 fluid (Walther and Orville, 1982). This type of behavior with respect to SiO_2 consumption has been found by Walther (1983) near the contacts between quartz veins and metacarbonate units.

From the discussion of this section, it appears that the fluid–rock ratios and metasomatic changes observed in carbonates may be explained by the single-pass fluid model provided carbonates constitute less than about 10% of the sequence. It is now appropriate to consider how such a model works with respect to the volumetrically more important pelites.

The Single-Pass Model and Fluid–Rock Ratio in Pelites

Phase equilibria and isotopic ratios in pelitic rocks are not, as discussed earlier, sensitive to fluid–rock ratio because the dominant fluid species produced during metamorphism are those that are close to equilibrium with metapelites. We will show, however, that even with single-pass flow fluid–rock ratios on the order of 6 : 1 can be recorded by metapelites in "ordinary" metamorphic terranes.

Let us begin by considering the implications of fluid–rock ratio calculations. Two possible end-members of fluid–rock ratio can easily be envisaged. If a rock produces all of the fluid involved in its metamorphism, the ratio of external fluid–rock is zero. On the other hand, any particular rock reacts within a metamorphic pile that is producing fluid for kilometers in all directions. If a rock equilibrates on the millimeter or centimeter scale with all of the fluid that passes through it, then the fluid–rock ratio approaches infinity. In other words, if all of the fluid that passes out of the top of the sequence has equilibrated with each millimeter thickness of rock in its path, then these thin rock elements could record near infinite fluid–rock ratios. As generally defined, however, the relationship between calculated fluid–rock ratio integrated through time and actual instantaneous fluid–rock ratio is that shown in Fig. 4. The instantaneous value must be much less than 0.01 under

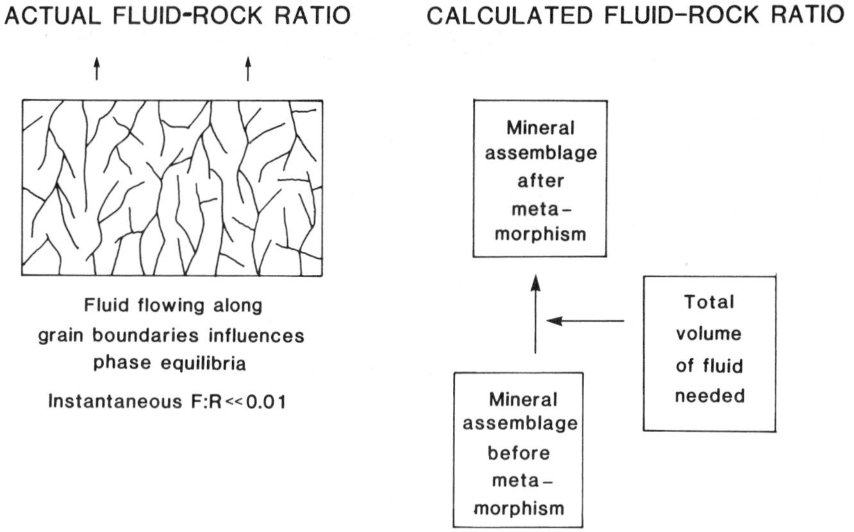

ACTUAL FLUID-ROCK RATIO   CALCULATED FLUID–ROCK RATIO

Fig. 4. A comparison of actual instantaneous fluid–rock ratio (left) and fluid–rock ratio calculated from mineral equilibria (right). Note that large apparent fluid–rock ratios will be calculated from some equilibria in the major rock types even with no addition of fluid from sources external to the metamorphic pile (see text).

most circumstances, but the former can, depending on the equilibria used to calculate it, be quite large. Since only the integrated effects are observed, the calculated fluid–rock ratio is to a large extent an artifact of the method of calculation and depends heavily on actual volumes of equilibration. While not entirely meaningless, integrated fluid–rock ratios must not force us to reach physically unrealistic conclusions about the nature of the metamorphic process. As an example of the calculation of high fluid–rock ratio in pelites, we will consider quartz precipitation.

Assume that a 6–10 km thick pile of quartz-saturated pelitic sediments is undergoing metamorphism in a geothermal gradient of 20–30°C/km. The pressure and temperature gradients are approximately 0.0028 bar/cm and 0.0002–0.0003°C/cm, respectively. SiO_2-saturated fluid flows upward throughout the pile, and because of the declining solubility of SiO_2 with decreasing pressure and temperature (Walther and Helgeson, 1977) quartz will precipitate as the fluid ascends. At each point in the pile, one can calculate the amount of silica precipitated per centimeter of vertical height per mole of passing fluid assuming that equilibrium is approached. The result is shown in Fig. 5 in terms of micromoles of SiO_2 per centimeter vertical height per mole of H_2O. It can be seen that the difference between 20 and 30°C/km gradients is quite small, and that the precipitation per mole of H_2O is greater at higher temperatures than at lower temperatures. Although, for convenience, the calculations were started at 5 kb and 600°C (Fig. 5), the

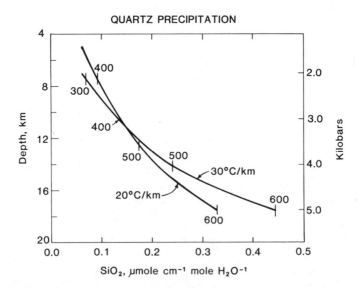

Fig. 5. Quartz precipitation accompanying upward migration of H_2O along two "typical" metamorphic geotherms (20°C/km and 30°C/km). Units are μmol SiO_2 per centimeter vertical height in the pile per mole of water passing.

starting point has little effect on the amounts of SiO_2 precipitated at lower temperatures.

In order to turn the rates of precipitation into a real fluid–rock ratio, we need a model metamorphic event. As illustrations, we will use events in which the rock sequence is dominantly pelitic and in which the temperature gradients are 20°C/km and 30°C/km. We treat this isobarically (rather than polybarically) and assume that the isotherms sweep upward through the metamorphic pile so that any particular rock sees most of its metamorphism isobarically in the temperature interval 400–600°C. In this temperature interval, the rates of quartz precipitation (P) in the rock per mole of H_2O per centimeter are given by isobaric curves similar to those of Fig. 5. Linear approximations corresponding to a pressure of 5 kb are (Walther and Helgeson, 1977):

$$\text{(30°C/km)} \quad P_{30} = \frac{0.4(T - 400)}{400} + 0.17 \ \mu\text{molSiO}_2/\text{cm}\cdot\text{mol H}_2\text{O} \quad \text{(2a)}$$

$$\text{(20°C/km)} \quad P_{20} = \frac{0.3(T - 400)}{400} + 0.12 \ \mu\text{molSiO}_2/\text{cm}\cdot\text{mol H}_2\text{O}, \quad \text{(2b)}$$

where T is temperature in degrees C.

Fluid is being released in the underlying rock pile (~2 mol/kg), and any particular rock element will experience its greatest flux when it is at 400°C and the underlying rocks are in the temperature range 400–600°C. When it reaches 600°C, the fluid flux (df_{H_2O}/dt) will drop to zero. Thus we have, for a rock that passes the 400°C isotherm:

$$\frac{df_{H_2O}}{dt} = \frac{2F}{\tau}\left(1 - \frac{t}{\tau}\right) \ \text{mol H}_2\text{O/cm}^2\cdot\text{s}, \quad \text{(3)}$$

where F is the total fluid released from the underlying rock column in the temperature range of 400–600°C and τ is the total time for the rock to heat from 400° to 600°C. If fluid release by reaction is a linear function of temperature, then $(T - 400)/400$ is equal to $t/2\tau$, and substituting (2a) into (3) gives, for 30°C/km:

$$\frac{dm}{dt} = \frac{2F}{\tau}\left(1 - \frac{t}{\tau}\right)\left|\frac{0.4t}{2\tau} + 0.17\right| \ \mu\text{mol SiO}_2/\text{cm}^3\cdot\text{s}.$$

A similar equation may be obtained for the 20°C/km geothermal gradient. Integrating to the time of interest enables solution for the total mass of silica deposited per cubic centimeter of rock. The solutions are shown graphically in Fig. 6 and indicate that on the order of 600 μmol of quartz would be deposited in each cubic centimeter of rock at peak temperatures of about 500°C. Volumetrically, this is an extremely small amount, only about 1.3% of the rock volume, and it would be undetectable unless, as often is the case, it formed coherent veins. The implications for fluid–rock ratio are great, however. Since the solubility of SiO_2 in H_2O at 500°C is on the order of 0.1 molal, each cubic centimeter of fluid contains 10^{-4} mol or 100 μmol of SiO_2.

ONE–PASS SYSTEM: SiO$_2$ PRECIPITATED AND FLUID–ROCK RATIO

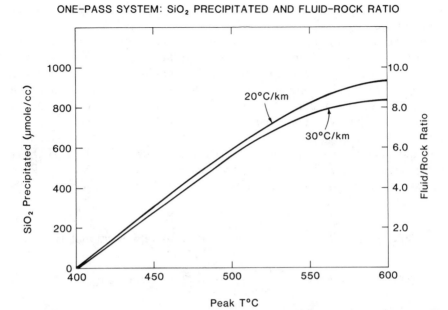

Fig. 6. Amounts of SiO$_2$ precipitated and calculated fluid–rock ratios for individual rock types during metamorphism of a 6–10 km pile of pelites at 20°C/km and 30°C/km. Fluid flow is of "single-pass" type. Note that high fluid–rock ratios can be obtained in a thin pelitic sequence without any external fluid source.

Addition of 600 μmol of SiO$_2$ per cubic centimeter of rock gives, unequivocally, a fluid–rock ratio of 6 : 1 from the conventional method of calculation. This fluid–rock ratio is a simple consequence of single-pass flow in a quartz-saturated pelitic sequence and does not require any external fluid or convection within the metamorphic pile. Even though each individual unit produces much less than 1 volume of fluid per volume of rock, single-pass flow of this fluid must, at equilibrium, yield a large fluid–rock ratio from the quartz equilibrium.

Any phase like quartz that dissolves congruently in the fluid may be treated in the same way. Most phases have incongruent dissolution relationships, however, and require the imposition of more specific compositional conditions in order to make the same kind of calculation. Nevertheless, it is instructive to calculate the effects of variable fluid–rock ratios on compositional or activity gradients in metamorphic rocks. Not only does this provide the basis for computation of apparent fluid–rock ratios in regions where demonstrable metasomatism has occurred, it also facilitates understanding of the reason why some activity gradients are preserved while others are destroyed. As examples, we will consider oxygen activity gradients and alkali and halogen activity gradients in metapelites.

Fluid–Rock Ratios and Oxygen Activity Gradients

Large gradients in oxygen activity are common features of interbedded meta-sediments that have been regionally metamorphosed in the amphibolite facies (Chinner, 1960; Rumble, 1973). Chinner (1960) reports interbedded pelites in the sillimanite zone of the Scottish Dalradian that contain he-matite, hematite + magnetite, magnetite + nearly pure ilmenite, and, occa-sionally, all three phases. Rumble (1973) found similar interlayered assem-blages in kyanite grade metaquartzites from western New Hampshire. Microprobe analyses of coexisting phases revealed oxygen activity differ-ences of about 4 orders of magnitude on the scale of a thin section (Rumble, 1973). Both authors concluded that the large oxygen fugacity gradients could only be consistent with oxygen being an inert, immobile component during regional metamorphism in these two areas. They suggested that the Fe^{2+}/Fe^{3+} ratios of the rocks were inherited from the sedimentary protoliths and had undergone little or no change during prograde metamorphism.

In contrast to these observations in regionally metamorphosed terranes, contact metamorphism often appears to produce substantial reduction of Fe^{3+}-bearing oxides to ferrous oxides and silicates. Chinner (1962) found that hematite and magnetite in the regionally metamorphosed rocks of Glen Clova had been reduced to produce Fe^{2+} silicates and hercynitic spinel where these entered the aureoles of the Lochnagar and Glen Doll intrusives. Similar observations may be made in the aureole of the Comrie diorite (Til-ley, 1924; Wood, unpublished observations) and in roof pendants of the Sierra Nevada batholith (Best and Weiss, 1964). Petrographic data force one to conclude that these contact aureoles have been "open" with respect to oxygen and that fluid derived from the intrusions or elsewhere has imposed the oxygen fugacity now recorded by the rocks. Stable isotope results on contact aureoles also confirm substantial oxygen exchange with externally derived fluid (Forester and Taylor, 1976). It is intriguing to note that, despite the large fluid fluxes attending regional metamorphism (thousands of moles per square centimeter), no effects on the oxides are observed whereas intru-sion at deep levels into almost anhydrous rocks (e.g., Lochnagar) has dra-matic results.

The differences between regional and "contact" events might, in part, be due to thermal convection and fluid recycling through the aureoles in the latter case and single-pass flow in the former. A simpler explanation that does not require such inferences about the nature of fluid flow may, how-ever, be found in the mineralogical expression of oxygen fugacities. In the case of intrusions such as Lochnagar and Comrie, the loss of opaque oxides and production of nearly pure $FeAl_2O_4$ spinel implies that the external fluid had an f_{O_2} close to the wustite–magnetite (WM) buffer (Turnock and Eug-ster, 1962). Consideration of equilibria such as:

$$3Fe_2O_3 + H_2 = 2Fe_3O_4 + H_2O$$

and

$$12Fe_2O_3 + CH_4 = 8Fe_3O_4 + CO_2 + 2H_2O,$$

enables calculation of the fluid–rock ratios required to reduce hematite to magnetite in H-O and C-H-O fluids, respectively. These calculations are shown for pure H_2O-H_2 and graphite-saturated fluids in Fig. 7. Fugacity and activity coefficients for fluid species were obtained from the MRK program of Flowers (1979), and the calculations were performed at 5 kb and 500°C. Considering a rock with about 5 vol% hematite, it can be seen from Fig. 7 that an H_2O-H_2 fluid at f_{O_2} 1 to 2 orders of magnitude above WM will readily reduce all of the hematite to magnetite with fluid–rock ratios of 0.1 to 1.0. Graphite-saturated fluids are even more efficient reducers and require fluid–rock ratios 2 orders of magnitude lower in order to effect complete reduction of hematite to magnetite. This is because of the predominance of CH_4 in the C-H-O fluid and the fact that 1 mol of CH_4 can reduce 12 mol of hematite. On the other hand, oxidation of magnetite to hematite is extremely difficult because of the small quantities of H_2 that can be produced in any fluid within the hematite stability field. For example, 1 order of magnitude above the

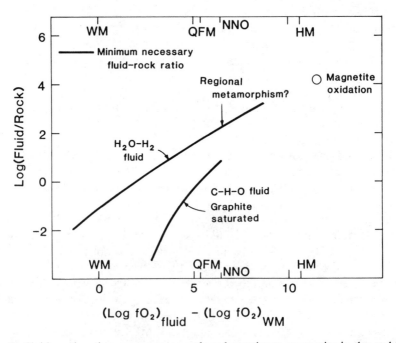

Fig. 7. Fluid–rock ratios necessary to reduce hematite to magnetite in the stability field of the latter. Note that hematite is readily reduced at low f_{O_2} particularly in graphite saturated fluids but that magnetite is very difficult to oxidize in the hematite stability field.

hematite–magnetite boundary fluid–rock ratios on the order of 10^4 (Fig. 7) would be required. Qualitatively, then, one can conclude that magnetite is very insensitive to the imposition of a high f_{O_2} fluid, whereas hematite is very readily reduced under conditions where hercynite spinel becomes stable. It appears, therefore, that small amounts of relatively reducing fluid can produce the substantial changes seen in many contact aureoles. That the fluids involved are often reducing is confirmed by the production of ferrous oxides and silicates.

In contrast to the locally imposed oxygen activities within thermal aureoles, the fluids flowing through regional metamorphic terranes represent averages of carbonate, pelite and psammite lithologies with corresponding average f_{O_2}'s. Average oxygen activity conditions in regional metamorphism of pelites would be expected, on the basis of Chinner's (1960) and Rumble's (1973) data, to be close to NNO. Under such conditions hematite would be quite resistant to reduction, even if the average fluid were graphite saturated. Magnetite could not be oxidized. Fig. 7 shows that fluid–rock ratios on the order of 5 to 110 would be required for complete reduction of hematite by a fluid at NNO in a rock containing 5 vol% of the oxide. Since the calculated fluid–rock ratio has an inexact relationship to real fluid–rock equilibria (Fig. 4), it is not clear whether or not these ratios could be routinely achieved in regional metamorphism. However, it is apparent that, given the assumption of average fluid near NNO, hematite-bearing rocks do not equilibrate with large volumes of externally derived fluid during regional metamorphism.

In summary, hematite reduction in deeply buried contact aureoles such as Lochnagar requires the introduction of less than 1 volume of external fluid per volume of rock. The maintenance of large activity gradients in interbedded regionally metamorphosed pelites, on the other hand, only constrains fluid–rock ratios to be less than about 10^2. Differences between "inert" and "mobile" oxygen behavior in regional and contact metamorphism may be explained, therefore, as due principally to differences in f_{O_2} of the infiltrating fluids involved. Large fluid–rock ratios are not necessarily required.

Activity Gradients in Micaceous Rocks

Micas are involved in a wide range of prograde and retrograde metamorphic reactions. They are, in principle, therefore, sensitive monitors of the natures of the fluid phases attending such transformations. Apart from differences in Al/Si and Fe^{2+}/Fe^{3+} ratio, these minerals exhibit substantial variations in the readily exchangeable cations Na^+ and K^+ and anions OH^-, F^-, and Cl^-. The ratios Na/K, F/OH, and Cl/OH depend on compositions of both fluid and solid phases and on temperature (e.g., Gunter and Eugster, 1980; Munoz, 1984). Since these species are readily exchangeable, however, the presence

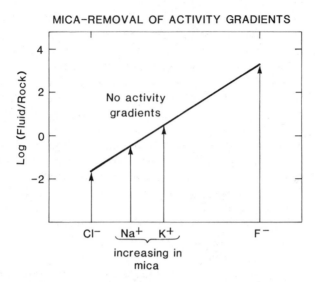

Fig. 8. Fluid–rock ratios necessary to remove activity gradients in Na^+, K^+, F^-, and Cl^- in rocks in which these species are dominantly contained in mica. Note that it is easier to make white mica more Na rich than more K rich (see text) and also that alkali activity gradients are commonly observed.

or absence of gradients in their activities constrain fluid–rock ratios recorded by micaceous rocks.

Detailed studies of fluid inclusions in metamorphic and pegmatitic rocks indicate that fluids attending metamorphic and hydrothermal processes have moderate salinities, with NaCl contents on the order of 5 wt% (Poty *et al.*, 1974). Both di- and trioctahedral micas have low affinities for Cl^-, however, and the data of Munoz and Swenson (1981) and Volfinger *et al.* (1985) indicate that trioctahedral micas are unlikely to contain more than 500 ppm Cl^- under these conditions of fluid composition. Therefore, a rock containing 10% trioctahedral mica should contain about 0.14 g of Cl per 1000 cm³ and coexist with a fluid containing about 30 g of Cl per 1000 cm³. Under these conditions a fluid–rock ratio of about 1:40 would be quite adequate to remove all gradients in Cl^- activity (Fig. 8).

To our knowledge, no detailed tests of Cl^- activity/concentration gradients have yet been made in metamorphic terranes. Such studies would require consideration of the effect of bulk composition on Cl/OH ratio in trioctahedral mica to be made. According to Munoz (1984), the affinity between Cl^- and Fe^{2+} yields a constant chloride activity parameter at variable Fe/Mg of:

$$C_{Cl} = -5.01 - 1.93 X_{Mg} - \log\left|\frac{X_{Cl}}{X_{OH}}\right|$$

where $X_{Mg} = \dfrac{Mg^{2+}}{Mg^{2+} + Fe^{2+}}$ and X_{Cl} and X_{OH} refer to atomic fractions of the subscript species on the hydroxyl site in the mica. The presence or absence of constant Cl^- activity recorded by micas in interbedded metamorphic rocks enable demonstration of minimum or maximum fluid–rock ratios in the range 0.01–0.1. These provide potentially important constraints on fluid–rock interactions during metamorphism.

In contrast to chloride, fluoride has a pronounced affinity for mica over aqueous fluid and, within the former, a preference for Mg-rich over Fe^{2+}-rich compositions. At constant fluoride activity, the activity parameter (Munoz, 1984) is:

$$C_F = 1.52X_{Mg} + 0.42X_{An} + 0.2X_{Sid} - \log \left| \frac{X_F}{X_{OH}} \right|, \qquad (4)$$

where $X_{Mg} = \dfrac{X_{Mg}}{X_{Mg} + X_{Fe}^{2+}}$. The siderophyllite component X_{Sid} is given by

$$X_{Sid} = [(3 - Si/Al)/1.75][1 - X_{Mg}],$$

and the annite component X_{An} by:

$$X_{An} = 1 - (X_{Mg} + X_{Sid}).$$

From their experimental OH-F exchange data, Munoz and Ludington (1974, 1977) obtained a relationship between fluid fugacities and the fluoride parameter C_F of:

$$\log(f_{H_2O}/f_{HF}) = \frac{2100}{T} + C_F. \qquad (5)$$

Guidotti's (1984) review of F^- contents of trioctahedral micas suggests an average wt% F of about 0.3% or $\dfrac{F}{F + OH}$ of approximately 0.03. This gives, at 500°C, from (4) and (5) a ratio of f_{H_2O}/f_{HF} of 120,000 for a mica of $X_{Mg} = 0.4$ and $X_{An} = 0.6$ coexisting with aqueous fluid. An MRK equation of state fitted to pressure–volume–temperature (*PVT*) data for pure HF gives fugacity coefficients similar to those for pure H_2O (J.R. Holloway, personal communication). Since the HF must be extremely dilute, it is necessary to take account of the apparent large activity coefficients (around 70) for dilute HF in H_2O-HF fluids (Westrich, 1978). Putting these data together suggests that a biotite with about 0.3 wt% F coexists with a fluid containing less than 1 ppm F^-. Therefore, in order to remove activity gradients in micaceous rocks, fluid–rock ratios in excess of 1000 : 1 by volume would be required (Fig. 8). This conclusion strongly supports the assertion that metamorphic micas generally retain the fluoride contents of the sedimentary protoliths and that F^- acts as an immobile component during metamorphism (Guidotti, 1984). This "immobility" does not reflect an inability for F^- to be exchanged but a severe mass-balance constraint on the removal of fluoride activity gradients.

Alkalis are intermediate between Cl^- and F^- in terms of ease of reequilibration with aqueous fluids. Gunter and Eugster (1980) give thermodynamic data for exchange equilibria such as:

$$NaAl_3Si_3O_{10}(OH)_2 + KCl = KAl_3Si_3O_{10}(OH)_2 + NaCl. \qquad (6)$$

     mica   fluid   mica   fluid

These enable calculation of equilibrium NaCl/KCl ratios in fluids coexisting with muscovite–paragonite micas at high pressures and temperatures. Taking the total salinity of metamorphic fluid to be on the order of 5 wt% and assuming around 30% white mica plus alkali feldspar in the rock, one can calculate the approximate fluid–rock ratio required to eliminate alkali activity gradients in the metasediments. This is done by taking the equilibrium constant for (6):

$$K = \frac{a^{mica}_{KAl_3Si_3O_{10}(OH)_2} \cdot a^{fluid}_{NaCl}}{a^{mica}_{NaAl_3Si_3O_{10}(OH)_2} \cdot a^{fluid}_{KCl}}$$

and converting to concentration using the activity-composition relations of Chatterjee and Froese (1975) and Gunter and Eugster (1980) for mica and fluid phases, respectively. Given the common range of white mica compositions (K/K + Na of 0.8 to 1.0), one can then calculate the mass of fluid of known K/K + Na that would have to be added to the system in order to force the micas all to move to essentially the same composition (within 1 or 2 mol%). Since the common white micas are K rich (K/Na ≥ 4) and the fluids coexisting with them are Na rich, the fluid–rock ratios required to remove activity gradients depend on whether the infiltrating fluids are relatively K rich or relatively Na rich. From Fig. 8 it can be seen that the low ratio of K to Na in the fluid makes it more difficult to force the micas to move toward K-rich compositions than Na-rich compositions. Thus, fluid–rock ratios on the order of 3 : 1 are required to even out activity gradients in mica if the infiltrating fluid is adding K to the rock. On the other hand, if the infiltrating fluid is adding Na and subtracting K, the high Na/K ratio of the fluid enables this to go to completion with a fluid–rock ratio of about 0.3 : 1. The nature of regional metamorphism is such that high-temperature fluids flow upward to lower temperatures exchanging alkalis en route (Fig. 1(b)). The temperature dependence of the equilibrium constant for (6) results, at constant bulk composition, in mica becoming more K rich and fluid more Na rich as temperature declines. Therefore, fluids flowing upward from high to low temperature add K to the mica and need fluid–rock ratios on the order of 3 : 1 to even out activity gradients.

Metasomatic changes that involve Na addition to the mica happen when fluid is flowing from low to high temperatures and would result in, for example, Na metasomatism such as that observed in ocean floor geothermal systems. Lower fluid–rock ratios are required in this case (Fig. 8).

Several studies have revealed wide ranges in Na/K ratios of micas and feldspars from both high-grade (Evans and Guidotti, 1966) and low-grade (e.g., Baltatzis and Wood, 1977) metapelites. It is generally thought (Evans and Guidotti, 1966; Guidotti, 1984) that the observed variations in alkali contents represent values inherited from the protolith and that little alkali metasomatism takes place during prograde metamorphism. The calculations illustrated here in Fig. 8 require there to be low apparent fluid–rock ratios in those cases where substantial alkali gradients occur.

Conclusions

Observations of fluid pressure in sedimentary basins reveal that fluid pressure becomes greater than hydrostatic at depths on the order of 3–6 km. Fluid cannot, therefore, circulate convectively below these depths and fluid released at depth can only flow upward towards the earth's surface in the "single-pass" manner depicted in Fig. 1(b). Since high-grade metasediments have passed through analogous diagenesis and lithification steps to those occurring in sedimentary-basin horizons, it seems that fluid circulation cannot generally occur at the depths of regional metamorphism. Fluid flow in regional metamorphism must, therefore, in general be of the single-pass type. Shallow-level contact phenomena may, on the other hand, involve circulation provided depths are shallow enough for the strengths of the rocks to hold open pores and fractures.

Application of the single-pass model to carbonate and pelite metamorphism leads to several conclusions. First, the fluid–rock ratios and fluid compositions observed in metacarbonates may readily be generated within the metamorphic pile if limestones occupy no more than 5–10% of the sequence by volume. This proportion of limestones is not an unusual one in metamorphic terranes. Second, despite the low volumes of H_2O generated during prograde metamorphism of pelites ($\sim$50 cm^3/380 cm^3 of rock) metapelites should, with respect to some equilibria, record very high fluid–rock ratios. This is illustrated by the deposition of quartz in a single-pass fluid-flow system. Assuming equilibrium crystallization in a typical geothermal gradient and taking regionally metamorphosed sequences of 6–10 km thickness, about 1 vol% of quartz should be deposited in an average pelite by upward-flowing fluid. This is equivalent to a fluid–rock volume ratio of 6 : 1 calculated in the conventional manner. Thus, the single-pass model readily produces high fluid–rock ratios in rocks that do not produce large amounts of fluid and that are the *dominant* components of the metamorphic sequence. It should be emphasized that these high fluid–rock ratios are internally generated in a relatively thin metamorphic sequence. Even higher values may obviously be attained if dehydrating sediment is being subducted below the metamorphic pile or if there is advection of hydrous magma from the underlying slab or mantle wedge.

Analyses of redox equilibria and exchange equilibria involving micas provide alternate means of constraining fluid–rock ratios. The observed reduction of hematite in contact aureoles (Chinner, 1962) can be effected with low fluid–rock ratios (<1 : 1) in both O-H and C-O-H fluids. The presence of large oxygen activity gradients in regionally metamorphosed rocks does not require low fluid–rock ratios. Magnetite is extremely insensitive to oxidation, and hematite reduction in the "average" terrane would require fluid–rock ratios on the order of 10^2. Therefore, the presence of large oxygen activity gradients in regional terranes and their destruction in contact aureoles does not require small fluid–rock ratios in the former and large ones in the latter.

Micas are potentially useful indices of fluid–rock interactions. For example, chloride activity gradients in interbedded rocks would be completely removed with fluid–rock ratios on the order of 1 : 40 whereas fluoride gradients remain until fluid–rock ratios approach $10^3 : 1$. Alkali exchanges between fluid and mica are intermediate in sensitivity, alkali gradients being removed by fluid–rock ratios on the order of 0.3 : 1 to 3 : 1. The commonly observed presence of large alkali gradients in interbedded metasediments imply that fluid–rock ratios do not, with respect to alkali exchange, approach these levels.

Acknowledgments

Financial support was provided in part by National Science Foundation grants EAR82-12502 and EAR83-18905.

References

Baltatzis, E., and Wood, B.J. (1977) The occurrence of paragonite in chloritoid schists from Stonehaven, Scotland. *Mineral. Mag.* **41**, 211–216.
Best, M.G., and Weiss, L.E. (1964) Mineralogical relations in some pelitic hornfelses from the southern Sierra Nevada, California. *Amer. Mineral.* **49**, 1240–1266.
Burnham, C.W. (1959) Contact metamorphism of magnesian limestones at Crestmore, California. *Geol. Soc. Amer. Bull.* **70**, 879–920.
Chatterjee, N.D., and Froese, E. (1975) A thermodynamic study of the pseudobinary join muscovite–paragonite in the system $KAlSi_3O_8$-$NaAlSi_3O_8$-SiO_2-H_2O. *Amer. Mineral.* **60**, 985–993.
Chinner, G.A. (1960) Pelitic gneisses with varying ferrous/ferric ratios from Glen Clova, Angus. *J. Petrol.* **1**, 178–217.
Chinner, G.A. (1962) Almandine in thermal aureoles. *J. Petrol.* **3**, 316–340.
Engelder, T. (1984) The role of pore water circulation during the deformation of foreland fold and thrust belts. *J. Geophys. Res.* **89**, 4319–4326.
Etheridge, M.A., Wall, V.J., and Vernon, R.H. (1983) The role of the fluid phase during regional metamorphism and deformation. *J. Metam. Geol.* **1**, 205–226.
Etheridge, M.A., Wall, V.J., Cox, S.F., and Vernon, R.H. (1984) High fluid pressures during regional metamorphism and deformation: Implications for mass transport and deformation mechanisms. *J. Geophys. Res.* **89**, 4344–4358.

Evans, B.W., and Guidotti, C.V. (1966) The sillimanite-potash feldspar isograd in western Maine, U.S.A. *Contrib. Mineral. Petrol.* **12**, 25–62.

Ferry, J.M. (1983) On the control of temperature, fluid composition, and reaction progress during metamorphism. *Amer. J. Sci.* **238-A**, 201–232.

Ferry, J.M. (1984) A biotite isograd in south-central Maine, U.S.A.: Mineral reactions, fluid transfer and heat transfer. *J. Petrol.* **25**, 871–893.

Flowers, G.C. (1979) Correction of Holloway's (1977) adaptation of the Modified Redlich–Kwong equation of state for calculation of the fugacities of molecular species in supercritical fluids of geologic interest. *Contrib. Mineral. Petrol.* **69**, 315–318.

Flowers, G.C., and Helgeson, H.C. (1983) Equilibrium and mass transfer during progressive metamorphism of siliceous dolomite. *Amer. J. Sci.* **283**, 230–286.

Forester, R.W., and Taylor, H.P. (1976) ^{18}O, D/H and $^{13}C/^{12}C$ studies of the tertiary igneous complex of Skye, Scotland. *Amer. J. Sci.* **277**, 136–177.

Fyfe, W.S., Price, N.J., and Thompson, A.B. (1978) *Fluids in the Earth's Crust.* Elsevier Publishing Co., Amsterdam. 383 pp.

Graham, C.M., Greig, K.M., Shepherd, S.M.F., and Turi, B. (1983) Genesis and mobility of the H_2O-CO_2 fluid phase during regional greenschist and epidote amphibolite facies metamorphism: A petrological and stable isotope study in the Scottish Dalradian. *J. Geol. Soc. London* **140**, 577–599.

Greenwood, H.J. (1975) Buffering of pore fluids by metamorphic reactions. *Amer. J. Sci.* **275**, 573–593.

Gregory, A.R., and Backus, M.M. (1980) Geopressured formation parameters, geothermal well, Brazoria County, Texas, in *Proceedings 4th U.S. Gulf Coast Geopressure-Geothermal Energy Conf.*, edited by M.H. Dorfman and W.L. Fisher. **1**, 235–311.

Guidotti, C.V. (1984) Micas in metamorphic rocks, in *Micas*, edited by S.W. Bailey. *Rev. Mineral.* **13**, 357–467.

Gunter, W.D., and Eugster, H.P. (1980) Mica-feldspar equilibria in supercritical alkali chloride solutions. *Contrib. Mineral. Petrol.* **75**, 235–250.

Magaritz, M., and Taylor, H.P. (1976). $^{18}O/^{16}O$ and D/H studies of igneous and sedimentary rocks along a 500 km traverse across the Coast Range Batholith into Central British Columbia at latitudes 54°–55°N. *Can. J. Earth Sci.* **13**, 1514–1530.

Munoz, J.L. (1984) F-OH and Cl-OH exchange in micas with applications to hydrothermal ore deposits, in *Micas*, edited by S.W. Bailey. *Rev. Mineral.* **13**, 469–493.

Munoz, J.L., and Ludington, S.D. (1974) Fluorine-hydroxyl exchange in biotite. *Amer. J. Sci.* **274**, 396–413.

Munoz, J.L., and Ludington, S.D. (1977) Fluoride-hydroxyl exchange in synthetic muscovite and its application to muscovite-biotite assemblages. *Amer. Mineral.* **62**, 304–308.

Munoz, J.L., and Swenson, A. (1981) Chloride-hydroxyl exchange in biotite and estimation of relative HCl/HF activities in hydrothermal fluids. *Econ. Geol.* **76**, 2212–2221.

Norton, D., and Knight, J. (1977) Transport phenomena in hydrothermal systems: Cooling plutons. *Amer. J. Sci.* **277**, 937–981.

Poty, B., Stalder, H., and Weisbrod, A. (1974) Fluid inclusion studies in quartz from fissures of western and central Alps. *Schweiz. Min. Pet. Mitt.* **54**, 717–752.

Rubey, W.W., and Hubbert, M.K. (1959) Role of fluid pressure in mechanics of overthrust faulting I. Mechanics of fluid-filled porous solids and its applications to overthrust faulting. *Geol. Soc. Amer. Bull.* **70**, 115–166.

108   B.J. Wood and J.V. Walther

Rumble, D., III (1973) Fe-Ti oxide minerals from regionally metamorphosed quartz-ites of western New Hampshire. *Contrib. Mineral. Petrol.* **42,** 181–195.

Rumble, D., III, Ferry, J.M., Hoering, T.C., and Boucot, A.J. (1982) Fluid flow during metamorphism at the Beaver Brook fossil locality. *Amer. J. Sci.* **282,** 866–919.

Thompson, J.B. (1959) Local equilibrium in metasomatic processes, in *Researches in Geochemistry,* edited by P.H. Abelson, pp. 428–457. Wiley, New York.

Tilley, C.E. (1924) Contact metamorphism in the Comrie area of the Perthshire Highlands. *Geol. Soc. London Quart. J.* **80,** 22–71.

Tilley, C.E. (1951) The zoned contact skarns of the Broadford area, Skye. *Mineral. Mag.* **29,** 621–666.

Turnock, A.C., and Eugster, H.P. (1962) Fe-Al oxides: Phase relationships below 1000°C. *J. Petrol.* **3,** 533–565.

Volfinger, M., Robert, J.-L., Vielzeuf, D., and Neiva, A.M.R. (1985) Structural control of the chlorine content of OH-bearing silicates (micas and amphiboles). *Geochim. Cosmochim. Acta* **49,** 23–36.

Walther, J.V. (1983) Description and interpretation of metasomatic phase relations at high pressures and temperatures. 2. Metasomatic reactions between quartz and dolomite at Campolungo, Switzerland. *Amer. J. Sci.* **283-A,** 459–485.

Walther, J.V., and Helgeson, H.C. (1977) Calculation of the thermodynamic proper-ties of aqueous silica and the solubility of quartz and its polymorphs at high pressures and temperatures. *Amer. J. Sci.* **277,** 1315–1351.

Walther, J.V., and Orville, P.M. (1982) Rates of metamorphism and volatile produc-tion and transport in regional metamorphism. *Contrib. Mineral. Petrol.* **79,** 252–257.

Walther, J.V., and Wood, B.J. (1984) Rate and mechanism in prograde metamor-phism. *Contrib. Mineral. Petrol.* **88,** 246–259.

Westrich, H.R. (1978) Fluoride-hydroxyl exchange equilibria in several hydrous minerals. Unpublished PhD. thesis, Arizona State University, Arizona.

Chapter 5
Fluid Migration and Veining in the Connemara Schists, Ireland

B.W.D. Yardley

Introduction

A knowledge of how fluids behave during regional metamorphism is central to any understanding of metamorphic processes, but studies of regional fluid migration are still in their infancy, and there is considerable controversy about the possible role of migrating fluids in metamorphism. There can be no doubt that fluids are driven off from many rock types during progressive metamorphism and that fluid can reenter rocks after the peak of metamorphism to give rise to retrograde changes. However, major uncertainties remain concerning the permeabilities and porosities of rocks undergoing metamorphism, and even the widespread assumption by metamorphic petrologists that fluid pressure (P_f) is approximately equal to lithostatic pressure (P_l) can seldom be independently verified.

It is a reflection of these uncertainties that recent models in the literature range from the view of metamorphosing rocks as essentially "wet," i.e., quite permeable with a pore fluid under high pressure that may often be convecting (Etheridge *et al.,* 1983) to the view that the rocks are essentially dry, with a distinct fluid phase present only intermittently, even though fugacities of volatile species may be quite high (Thompson, 1983). Detailed studies of specific rock layers have successfully demonstrated that extensive mineralogical and chemical changes may result from fluid movement, and imply large fluid–rock ratios. This has been clearly demonstrated, for example, by the work of Ferry on the calc-silicate bands intercalated with more argillaceous layers in the Vassalboro Formation of Maine (Ferry, 1976, 1980, 1983) and by the work of Rumble *et al.* (1982) and Tracy *et al.* (1983). Even larger fluid–rock ratios (<20:1) were reported by Graham *et al.* (1983) to have resulted from interactions between fluids derived from interbedded metasedimentary and metavolcanic units in southwest Scotland.

Despite this carefully documented evidence for high fluid–rock ratios in specific instances, most studies have been of originally carbonate-bearing lithologies, and therefore their results may not be typical of the metamorphic pile as a whole. Stable isotope studies by Rumble and Spear (1983) demonstrate examples both of extensive fluid movement leading to isotopic equilibration with introduced fluid and of very local disequilibrium implying only a small fluid flux.

In this chapter, I briefly review some of the factors that control fluid pressures and fluid flows during metamorphism and attempt to interpret some of the veining phenomena observed in the Connemara Schists of Ireland in the light of these considerations.

Fluid Flow, Permeability, and Fluid Pressure

The flow of fluid through any rock is a result of deviations of fluid pressure from the stable hydrostatic vertical gradient, and the magnitude of the flow in any rock unit depends on its permeability. Brace (1980) has shown that for crystalline rocks in the upper part of the crust, permeabilities are sufficiently large that fluid pressures much greater than hydrostatic cannot be retained over geological periods of time.

The permeability of a rock unit as a whole may either reflect the ease with which fluid can move through it pervasively along microcracks, pores, or grain boundaries, or it may be a measure of movement along more widely spaced fractures. The compilation by Brace (1980) demonstrates that the "laboratory" permeability of most rock types, which is measured on small samples and is a measure of pervasive flow through the bulk rock, is usually several orders of magnitude less than the "in situ" permeability, measured on a larger scale within the crust (e.g., down wells), which is dominated by fluid flow along spaced fractures. Hence, the distinction that will be made in this paper is between the pervasive (or laboratory) permeability, irrespective of whether it is due to interconnected pores or to microcracks on the scale of the grain size, and fracture (or in situ) permeability, dominated by flow along spaced fractures on a scale of at least tens of centimeters.

Laboratory permeabilities are strongly dependent on the effective pressure, P_e (where $P_e = P_l - P_f$), according to Brace $et\ al.$ (1968) and Fyfe $et\ al.$ (1978). This is interpreted as reflecting the fact that abundant microcracks may be held open by the high fluid pressure when P_e is small, but only the strongest, spherical or tubular, pores can remain fluid filled if P_e is large, under metamorphic conditions (Brace, 1980).

Permeability increases markedly as the effective pressure declines and probably changes very rapidly when P_f approaches P_l. In addition, since devolatilization reactions lead to a decrease in the volume of the solid phases present, there may be a transient increase in porosity while reactions are taking place that will also enhance permeability (Fyfe $et\ al.$, 1978; Rumble $et\ al.$, 1982). Both Rye $et\ al.$ (1976) and Rumble $et\ al.$ (1982) report a correlation between the extent of oxygen isotopic exchange in individual carbonate rock samples, and the amount of CO_2 that had been released from the sample during metamorphism.

When the fundamental relationships between fluid flow, permeability, and fluid pressure are reviewed in the light of our understanding of metamorphism, a paradox emerges. If, as there is abundant evidence to suggest (e.g., Etheridge et al., 1983, 1984), fluid pressures are high during metamorphism, then permeabilities must be very small in order to sustain $P_f \gg P_h$, where P_h is the fluid pressure that would be present because of a hydrostatic head alone. However, the extensive flow of fluid through specific layers that has been documented in a number of instances clearly requires a high permeability. Part of the answer to this paradox probably lies in the production of fluid as reactions proceed, and a conceptual model for metamorphic fluid loss that takes this into account is developed below.

Fracturing of Metamorphic Rocks at High Fluid Pressure

Higher fluid pressures lead to an increase in the permeability of any rock because they facilitate the opening of microcracks by reducing the effective compressive stresses acting across them (Brace, 1980). If, however, the fluid pressure continues to rise despite this enhancement of permeability, the rock will ultimately fracture. This may take the form of either extensional hydraulic fracture (Secor, 1968) or shear failure.

In an isotropic medium, hydraulic fracturing will occur normal to the least principal compressive stress, σ_3, when P_f exceeds σ_3 by an amount greater than the tensile strength of the rock. Extensional veins should therefore form perpendicular to σ_3, as indicated by lineations and strain markers. In many rocks, however, the most abundant extensional veins are more or less parallel to schistosity, and this probably reflects a marked variation in the tensile strength with orientation in metamorphic rocks.

If a deviatoric stress is present, increased fluid pressure may lead to shear failure (Fig. 1). In this case, fracturing may occur when $P_f < P_l$ and the permeability of the bulk rock is still relatively low. Examples of quartz veins in both extensional and shear fracturing are found in the Connemara Schists and are illustrated below.

Influence of Deformation and Recrystallization on Permeability

Experimental studies by Brace and Orange (1968) demonstrated that deformation under brittle conditions causes an increase in permeability, resulting from dilatancy, in most rock types. However, even at low temperatures marble showed a reduction in permeability under stress, and this was interpreted by Brace (1968) as resulting from mineral plasticity that allowed

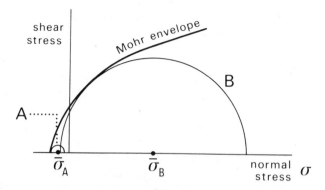

Fig. 1. Mohr circle diagram illustrating a schematic failure envelope for a metamorphic rock. Semicircles A and B represent the stress fields for the cases of small and large deviatoric stress and illustrate extensional failure with the mean stress tensional (e.g., $P_f > P_l$) and shear failure with a compressive mean stress ($P_f < P_l$), respectively.

cracks to seal up; he suggested it might provide a model for the behavior of silicate rocks at elevated temperatures. This implies that rocks which deform during metamorphism may undergo a reduction in permeability that inhibits fluid escape by flow through the bulk rock and leads to the formation of discrete fractures. If this mechanism is an important cause of fracturing, we would predict a correlation between veining events and episodes of deformation.

Fluid Production and Release during Prograde Metamorphism

We can develop a simple conceptual model for fluid release by taking as the starting point a deeply buried sediment containing pore fluid, but with $P_f < P_l$ and a correspondingly low permeability. Spaced cracks are absent. On heating, fluid is released by mineralogical reactions causing a rise in both P_f and permeability, irrespective of whether spaced fractures develop. Once the permeability has increased to the point where the rate of fluid loss balances the rate of fluid production, a steady state is attained.

Many metamorphic rocks are extensively veined, and the veins probably represent fractures along which fluid flow was concentrated. The development of fracture systems implies that the permeability of the unfractured rock was too low for fluid to escape at the rate it was being produced. As a result, P_f rose until either shear failure or extensional hydraulic fracture took place, thereby increasing the permeability of the bulk layer and focusing fluid flow into spaced cracks. Once fluid flow is concentrated into a fracture, then a quartz vein will inevitably be produced if the rocks are siliceous and flow is from hotter to colder regions (Walther and Helgeson, 1977). In other circumstances, a conspicuous mineral vein may not develop and the former fracture may go unnoticed.

On the other hand, in some metamorphic terranes veins are extremely scarce. Yardley (1983) noted that the low-grade slates of the Harlech Dome, Wales, are largely unveined, as are many thermal aureole rocks. Rumble *et al.* (1982) reported that impure fossiliferous marbles had undergone extensive devolatilization with a high fluid–rock ratio of the order of 4 : 1, without any fracturing being apparent. In these cases, the pervasive permeability of the rock was clearly sufficient to permit fluid to escape while P_f was below the threshold for fracturing.

The critical factors that determine whether or not veins form in a particular layer are therefore the rate of fluid production or influx, and the pervasive permeability of the rock. We have seen above that pervasive permeability is diminished in the case of rocks undergoing deformation at high grades, and the style of any fractures that do develop during deformation will depend on the magnitude of the deviatoric stress. Hence, rocks that are undergoing deformation as reactions proceed are the most susceptible to veining according to this criterion, and this hypothesis is tested against field observations below. Equally, a rapid rate of fluid release will tend to promote fracturing, although this factor may not be very important for most instances of veining, since many thermal aureoles whose rocks must have been rapidly heated lack veins.

The Rate of Fluid Production

In general, the strongly endothermic character of devolatilization reactions means that reaction rate must be controlled by the supply of heat (Yardley, 1977a; Fisher, 1978; Walther and Orville, 1982). The principal exceptions to this generalization probably occur where aqueous fluids infiltrate carbonate horizons and are unlikely to affect more than a small part of the metamorphosing pile.

The heat input to a succession of rocks largely passes through it, maintaining the steady-state temperature gradient. If, however, the heat input exceeds that required to maintain the existing temperature, the additional heat will be consumed in raising T or driving endothermic reactions. Walther and Orville (1982) calculated the rate of fluid production for a given difference in heat flow between bottom and top of a metamorphosing pile, and for this paper I have performed a similar calculation assuming that there is a decrease in heat flow within the pile of 40 mW m^{-2}. Taking the mean ΔH_r for devolatilization reactions as 60 kJ per mol H_2O liberated, and assuming constant T, these figures yield a time t to generate 1 mol H_2O per m^2 (normal to the direction of heat flow) of 1.5×10^6 s. Assuming negligible porosity, so that all the fluid generated must flow out of the rock, there will be a corresponding fluid flux of 6.7×10^{-7} mol m^{-2}s^{-1}, or if the molar volume of the aqueous fluid is taken to be 2×10^{-5}m^3, the flux, J_f, can be expressed as 1.33×10^{-11}m s^{-1}. This is similar to the figure obtained by Walther and Orville (1982) and probably represents an upper limit, perhaps more appropriate to a thermal aureole. It is unlikely, however, to be more than 2 or 3 orders of magnitude too high for most types of regional metamorphism.

Fluid flux is a function of hydraulic conductivity (K) and the hydraulic gradient driving the flow ($\Delta h/\Delta l$):

$$J_f = -K \cdot \Delta h/\Delta l, \tag{1}$$

where Δh is a notional hydraulic head driving the flow and Δl the distance across which it operates.

If P_e is constant throughout the system so that the gradient in fluid pressure is equal to the lithostatic pressure gradient, then even though P_f may be less than P_l:

$$\Delta h/\Delta l = \rho_r - \rho_f, \tag{2}$$

where ρ_r and ρ_f are the densities of rock and fluid, respectively, and are taken to be 2.8 and 0.9. Substituting in 1 yields:

$$K = 6.8 \times 10^{-12}\text{m s}^{-1}.$$

Hydraulic conductivity depends on both rock and fluid properties but is related to permeability (k), an exclusively rock property, by:

$$k = \nu K/g, \tag{3}$$

where ν is kinematic viscosity and g is the gravitational constant. Taking an approximate value of ν for amphibolite facies metamorphism as 1.5×10^{-7} m^2 s^{-1} (extrapolated from Norton and Knight, 1977) yields:

$$k \simeq 10^{-19} \text{ m}^2.$$

This calculation yields a minimum value for the in situ permeability of a layer undergoing metamorphism to permit the fluid driven off to escape and is of course dependent on the precise reaction rate. The implication of the field observations noted above is that in some instances the pervasive or laboratory permeability of the rock attains a value of this magnitude, but in others the laboratory permeability is less than this value when P_f rises to the point where fracturing occurs. If the consequence of fracturing is to give rise to a greater permeability than 10^{-19} m^2, then either fluid pressure will drop or the fractured layer will act as a focus for the influx of fluid from adjacent horizons, or both.

In the following section, the detailed field relations of veins in the Connemara Schists are described, with a view to exploring the importance of fracturing for metamorphic fluid release and the controls on when it occurred.

Evidence for Fluid Movement in the Connemara Schists

Regional Setting and Metamorphic History

The Connemara Schist belt comprises a sequence of Eocambrian to Cambrian metasediments including quartzites, pelites, marbles, and psammites, with horizons of basic metavolcanics (Fig. 2). The succession is correlated with the Dalradian Supergroup of the Scottish Highlands. Metamorphism during early to middle Ordovician times followed a complex $P-T$ path (Yardley *et al.*, 1980). Initial metamorphism was of Barrovian type, comparable to that of the Dalradian elsewhere in Ireland, but the early kyanite zone assemblages were subsequently overprinted over much of the area by a low-pressure, high-temperature metamorphism that appears to be the result of the emplacement of a sequence of calc-alkaline magmas ranging from gabbros through voluminous quartz diorite to granite that are exposed in the southern part of Connemara (Leake, 1969). The isograds that have been mapped in the region (e.g., Yardley *et al.*, 1980) invariably relate to the later, low-pressure overprint and have a simple east–west trend. Isograds relate to the appearance of sillimanite and higher grade reactions culminating in extensive partial melting; the lowest grade, staurolite–chlorite zone, rocks have been little affected by the low-pressure event. Despite the significant pressure difference (>3 kb) between the Barrovian event and the later metamorphism in which cordierite and andalusite, as well as sillimanite, were developed on a regional scale, they appear to be closely related in time and to represent a continuum. In the northern part of Connemara, with which this paper is primarily concerned, staurolite and rare kyanite grew after a major phase of isoclinal folding (D_2 in local terminology). The early basic phase of the calc-alkaline magmatism seen to the south was probably already initiated, or even completed, by this point, at least in terms of a chronology developed relative to structural phases. The entire sequence of metasediments and pre- to post-D_2 intrusives were then refolded into major D_3 folds that comprise northward-facing nappes in the northern half of Connemara. Such structures do not appear to be developed in Dalradian rocks elsewhere. Calc-alkaline magmatism continued to at least syn-D_3, and the more granitic phases may have continued beyond this time.

Metasediments close to the calc-alkaline intrusives are extensively migmatized (Leake, 1981), and high-temperature metamorphism in this southern, high-grade region clearly predates D_3 folding because sillimanite *faserkiesel* in sillimanite–K-feldspar quartzites have ellipsoidal forms that mirror the variation in D_3 strain around minor folds. High temperatures evidently continued during D_3 because in other lithologies sillimanite mats and migmatite leucosomes may be found aligned in S_3. Farther north, in the staurolite–sillimanite transition zone, sillimanite growth is locally preceded by the formation of coarse prismatic andalusite that is often partially re-

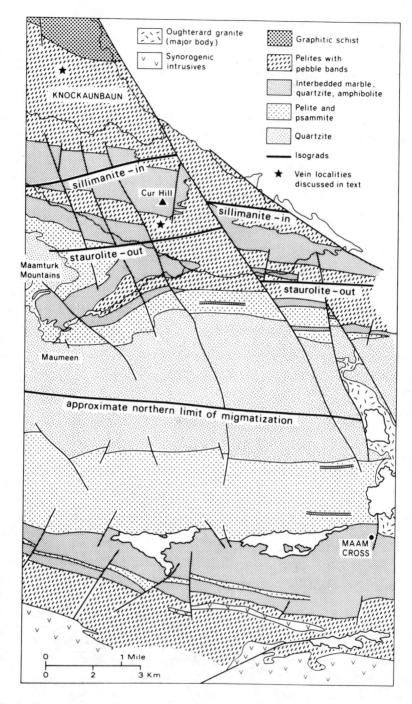

Fig. 2. Sketch map to illustrate the geology of part of eastern Connemara, Ireland, after Leake (1981), with isograds (Yardley *et al.,* 1980; Barber and Yardley, 1985) and localities discussed in sketch.

placed by fibrolite. Most of the andalusite occurrences are in veins (Fig. 3(a)), but coarse porphyroblasts sometimes occur in the pelites; they are, however, usually restricted to outcrops where andalusite veins also occur. Biotite inclusion trails in andalusite porphyroblasts suggest that their growth overlapped the early stages of the D_3 deformation, and so peak temperatures with the growth of sillimanite were only attained during D_3 in this region. The detailed sequence of metamorphism and deformation across the area is summarized in Fig. 4. It is fundamental to interpreting the veining phenomena because these can also be dated relative to structures.

Fluid Movement during Metamorphism

(1) *Mineralogical Constraints.* Chemical analyses suggest that pelitic schists in Connemara have not in general undergone significant metasomatism during metamorphism up to the onset of melting, although the highest grade pelites may be somewhat depleted in SiO_2 (Yardley, 1977b). Graphite is frequently absent from the pelites, or restricted to fine inclusions in feldspar, and so the fluid phase is assumed to have been H_2O dominated. This is in agreement with observations on fluid inclusions (Yardley *et al.,* 1983; Yardley, 1983).

In contrast, the peak metamorphic assemblages in the marbles appear to imply an internally buffered, H_2O-poor fluid. For example, Leake *et al.* (1975) report widespread serpentine pseudomorphs after olivine from the dolomitic Connemara Marble formation. Subsequent low-T hydration permitted the development of an ornamental marble industry but unfortunately made the study of peak assemblages very difficult. Dolomite, calcite, and diopside are, however, widespread. Marbles on Fig. 2, from the younger Lakes Marble Formation, are interbedded with pelitic schists and are dominantly composed of calcite. They had a peak assemblage of calcite + quartz + diopside + bytownite. Again, the development of retrograde phases, including pumpellyite, prehnite, zoisite, albite, and microcline, often makes it difficult to determine which minerals were present at peak temperatures. Nevertheless, despite their being interbedded with pelites, there is no evidence for influx of water into the marbles at the peak of metamorphism. This may result from the extremely ductile behavior of carbonate-dominated marbles at these temperatures, which would tend to keep their permeability to a minimum. The only evidence for fluids moving between pelites and marbles at peak temperatures comes from fluid inclusion studies of quartz veins in both lithologies (Yardley, 1983), indicating that such flow was concentrated in spaced cracks, and did not affect the bulk rock on each side. Conversely, however, there was extensive and more or less pervasive infiltration of water into marbles along cracks during uplift and cooling.

(2) *Veining Phenomena.* Quartz veins are universally present in the pelitic schists but vary in their thickness, abundance and age; they also occur less commonly in all other rock types and are locally abundant in some quartz-

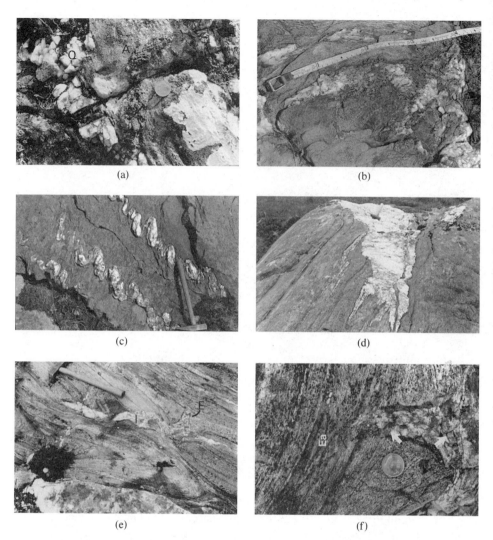

Fig. 3. Field photographs of veins in the Connemara Schists. (a) Andalusite vein (A) with negligible quartz, adjacent to a quartz vein (Q). (b) Quartz vein with marginal metasomatic effects including growth of prismatic andalusite, Knockaunbaun (see also Fig. 6). (c) Pre-D₃ vein in quartzite, folded in D₃, Maamturk mountains. (d) Large vein with apophyses, forming part of a major vein body, south of Cur Hill. (e) Quartz vein in internal boudinage structure (I), cutting extensional quartz vein (E). (f) Detail of the extensional vein in (e) to show it passing into an andalusite vein (Ea) in the adjacent pelite layer. Only minor andalusite (arrowed) is present where it cuts the quartz-rich horizon, which it does at a high angle to the D₃ stretching direction.

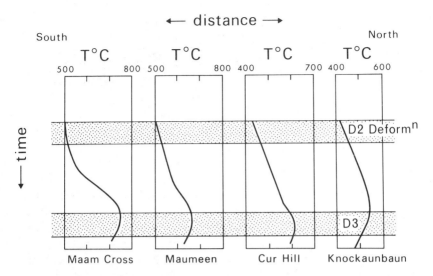

Fig. 4. Schematic representation of the variation in metamorphic temperature relative to major deformation events, across eastern Connemara.

ites. Veins are usually rare in amphibolites, while in marbles the bodies of vein quartz are lenses seldom exceeding a few tens of centimeters in length, and do not form continuous veins. Quartz is the dominant and usually the only vein mineral; however, tourmaline is a common accessory and plagioclase is occasionally present. Coarse andalusite prisms, 1–2 cm across and up to 10 cm in length, are found in some veins in the pelitic schists and usually have muscovite and quartz associated with them. Some andalusite veins are almost devoid of quartz (Fig. 3(a)), while in others, quartz may dominate. One distinct class of andalusite veins is present only in the northern part of Knockaunbaun townland (Fig. 2) and is in lower grade rocks than other andalusite veins. The andalusite here is mostly developed in the wall rocks to quartz veins, where it forms distinct "coralliform" poikiloblasts (Fig. 3(b)). The Knockaunbaun veins formed late in the metamorphic and structural sequence and are apparently related to retrograde changes. They are described separately below.

In the majority of cases, quartz veins show little evidence for an origin by local segregation from the immediate wall rocks. However, pelitic vein walls are sometimes depleted in quartz and feldspar for about 1 cm from the vein, and in the case of thin veins these depleted zones may provide an adequate source for the vein to have grown by a local segregation process, without significant infiltration (Yardley, 1975).

Relative Ages of Veins

Most veins are more or less parallel to the regional schistosity, which is a composite fabric, produced by the combined effects of D_2 and D_3 deformations, and is itself more or less parallel to bedding. Where folds are present, relative ages of veins are readily apparent, while elsewhere many veins exhibit the effects of specific deformations and may also be dated.

The most conspicuous small-scale folds in much of Connemara are D_3 folds, and pre-D_3 veins are often boudinaged or rodded parallel to D_3 fold axes. Some pre-D_3 veins appear to be related to MP_2 staurolite-grade metamorphism and have margins rich in somewhat coarse staurolite. These veins often form lenses up to 10 cm across. Veins may also be folded by D_3 folds (Fig. 3(c)). In some cases, earlier, pre-D_2 veins have been identified, but these are often difficult to distinguish. Very early (pre-D_2) veins are typically thin, 1–2 cm across, but may have been smeared out and disrupted by subsequent deformation.

Veins that are entirely post-D_3 in age are also readily identified and occur in most rock types. They are laterally continuous for tens of meters in some instances but are usually only a few centimeters in width. Characteristically, they are straight, parallel-sided, and cross-cut foliation; however, the late veins at Knockaunbaun have a less regular form.

The most abundant veins in northeast Connemara show complex relations to the D_3 structures. In part they are cross-cutting, but they also often exhibit boudinage or rodding; they are interpreted as each having formed within a relatively short time span during the D_3 event. In the pelites south of Cur Hill (Fig. 2), individual veins of this type often occur intimately associated with one another in composite veins, up to 2 m across, that can be traced along strike for several hundred meters (Fig. 3(d)). South of Cur Hill, these major vein sets are repeated with a 20–50 m spacing and are associated with thin siliceous beds. Individual quartz veins within the composites are often less than 10 cm in width but range up to large masses 1 m across and 12 m in length. Most individual veins are parallel to the overall trend of the composite veins, themselves approximately parallel to D_3 fold axial surfaces; however, short extensional veins occur at right angles to this trend and perpendicular to the D_3 stretching lineation (Fig. 3(e)). These veins are also interpreted as syn-D_3, with their orientation controlled by the regional stresses rather than dictated by the anisotropy of the host rocks.

Some syn-D_3 veins occur in "internal" or "foliation boudinage" structures (Cobbold et al., 1971; Platt and Vissers, 1980; Hambray and Milnes, 1975) rather than being simply extensional. Typically, such veins are concordant with the S_3 foliation on one side, but because of discordance of the foliation across the vein they are cross-cutting on the other side. A typical example is illustrated in Figs. 3(e) and 3(f), and in this case it cross-cuts an earlier vein formed by extension in the D_3 stretching direction. It seems reasonable to conclude that both veins are syn-D_3, and each formed separately within a short period of time relative to the duration of the deformation event. Note that the earlier vein is composed of quartz with minor andalusite where it cross-cuts a quartzitic layer but passes directly into an andalusite vein in the adjacent pelite, and this has been deformed during D_3 by being rotated toward parallelism with S_3.

Summary of the Field Observations on Veins

Quartz veins are found throughout the Connemara Schists and have formed at many different times. However, they are abundant and form large bodies only in predominantly pelitic units or, locally, in quartzite. Many veins are minor features of the order of 1–2 cm in width and may be essentially segregation features, but major veins displaying little evidence for local segregation are locally abundant, especially in pelitic schists near the sillimanite isograd. The largest vein sets formed mostly in a single episode from pre- to syn-D_3; however, each individual vein formed over a short part of this time span only. As well as being approximately coincident with D_3 ductile deformation, this major veining episode also corresponds with heating to peak metamorphic temperatures.

The Origin and Significance of the Major Quartz Vein Sets

(1) *The Causes of Fracturing.* In the first part of this paper it was estimated that, even with rapid heating, rocks usually need have a permeability of only around 10^{-19} m^2 for fluid to escape as fast as it is produced, and that in some areas this condition appears to have been attained without fracturing or veining.

A range of possible causes for fracturing and veining was also considered, and their relative importance in the Connemara Schists can now be evaluated in the light of the field evidence.

The most extensive and largest veins formed relatively late in the metamorphic heating, after most of the original volatile content of the rocks had already been driven off. While many small early veins may have been obliterated during subsequent deformation, it seems improbable that there could have been large early veins, on the scale developed during amphibolite facies metamorphism, present initially and subsequently destroyed. Partly this must reflect the increased solubility of quartz at high temperatures, but there is no correlation between amount of fluid released and amount of veining; nor is there any apparent mechanism for rates of devolatilization to have been enhanced to produce veins. The most extensive veins formed during deformation and are best developed in those rocks near the sillimanite isograd that were undergoing reactions to produce Al-silicate at the time that the D_3 deformation was taking place. Rocks now at higher grades, which must have undergone the same reactions prior to D_3 folding, did not develop comparable vein sets. This confirms that veining is not a response to a particular metamorphic reaction, but nor does it inevitably accompany a particular deformation episode. Rather, veins developed when a period of fluid release coincided with a deformation episode. In the absence of devolatilization reactions, the reduction in permeability during syntectonic recrystallization is of no great consequence, but when it prevents escape of fluids that are being internally generated, the resulting rise in fluid pressure may lead to cracking. From Fig. 1 it is apparent that if the deviatoric stresses are large during deformation, the cracks will be shear failure planes, whereas if they are small, extensional hydraulic fractures will dominate. In the field, veins can be found occupying both extensional and shear fractures (Fig. 3), but the extensional veins dominate. From this it is concluded that, even during deformation, deviatoric stresses were often sufficiently small to permit fluid pressure to rise to the point where hydraulic fracture occurred. In most cases this fracturing was controlled by the rock fabric, but in some instances it was normal to the least principal compression. Not uncommonly, however, foliation boudinage veins were formed, and these probably result from shear failure at $P_f < P_l$ owing to the presence of the deviatoric stress. Hence, the dominant cause of veining appears to have been reduction of bulk rock permeability during syntectonic recrystallization, as a result of which fluid pressure rose to exceed lithostatic. It is not clear whether the frequent association of major veins with siliceous beds in the pelite is merely the result of competency contrasts controlling the site of fracturing, or whether such contrasts were an essential factor causing the fractures to form.

(2) *The Amount of Fluid Involved in Prograde Veining.* Although veins are widespread, there is a particular area south of Cur Hill (Fig. 2) in which they make up at least 1% of the surface rocks and are typically large and lacking any evidence for quartz depletion of the wall rock that might indicate a local segregation origin. If we make the assumption that the quartz was precipitated in response to cooling of a migrating fluid, then it becomes possible to calculate the amount of fluid needed to account for the volume of the vein quartz, *although this calculation is entirely meaningless should the assumption prove false.*

Taking a temperature of precipitation of 575° and a maximum possible value for the temperature at source of 650° (limited by the onset of melting) then, for $P = 4$ kb (Barber and Yardley, 1985), the change in quartz solubility is approximately 0.11 m from the compilation of Walther and Helgeson (1977) with some further extrapolation. This value corresponds to precipitation of 6.73 quartz by 1000 g fluid or a fluid–quartz ratio of 150:1 by weight. If vein quartz makes up 1% of the outcrop, this is equivalent to a fluid–rock ratio of 1.5:1 when integrated over the entire pelite horizon.

The question of how such high apparent fluid–rock ratios can be attained, especially throughout major stratigraphic units, is one of considerable interest at present. Here, three types of mechanism will be considered: one-way flow of fluid channelled into particular layers, recycling mechanisms whereby the same fluid is reused as an agent for solution-redeposition, and diffusional processes that do not in fact involve flow.

Fluid Recirculation Mechanisms

The same fluid may be reused during veining either as a result of convective circulation or by pumping in the vicinity of propogating fractures.

Convective circulation was rejected by Walther and Orville (1982) as a possible process during regional metamorphism because it requires a hydrostatic gradient in fluid pressure, whereas other evidence suggests near-lithostatic fluid pressures. On the other hand, Etheridge *et al.* (1983) suggest the existence of convection cells below "cap-rocks" that permit $P_f = P_l$ at their base. The main limitation to the scale on which such cells could develop is provided by rock strength and is illustrated in Fig. 5(a).

Suppose that at a point X at the base of a cap rock:

$$P_f^X = P_l^X = P,$$

then at a point Y, a distance h below X:

$$P_f^Y = P + \rho_f gh, \ P_l^Y = P + \rho_r gh, \ P_e^Y = (\rho_r - \rho_f)gh.$$

Where P_e exceeds rock strength, open fluid-supported fissures will close up, rock permeability will decline, and convective circulation will no longer be possible. Hence, rock strengths control the maximum depth of a convective circulation cell, and for low strengths of a few bars or tens of bars, restrict it to a few tens of meters below the cap rock. Such cells could not, therefore, leach quartz from deep rocks at significantly higher temperatures than those at the top of the cell. On the other hand, even small and transient cells, capped by marble or amphibolite layers, could result in quartz veining if the driving force for the solution-redeposition process was not a temperature gradient but the free energy difference between the strained quartz grains in the bulk rock and the newly forming, and hence strain free, grains growing in a vein (Yardley, 1975). Additionally, fluid pressure differences between fissures and wall rock pore fluid would provide a small chemical potential gradient (Bruton and Helgeson, 1983). Although there is no reason to suppose that these subtle influences on chemical potential are likely to appear significant by comparison with the effects of T and P, the occurrence of local segregation features, such as that illustrated by Yardley (1975), demonstrates that they may cause appreciable changes in rocks nonetheless.

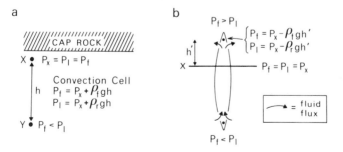

Fig. 5. Possible variations in P_f, P_l, and P_e in a hypothetical fluid-filled crack and its adjacent walls. (a) To illustrate changes below a hypothetical cap rock with $P_f = P_l$ at its base. (b) To demonstrate variations around an advancing fissure with $P_f = P_l$ at some arbitrary level along its length.

An alternative model that involves reusing the same fluid to attain higher fluid–rock ratios is illustrated in Fig. 5(b). Here, fluid movement is assumed to be sporadic rather than continuous within any one small area and involves the passage of successive "packets" of fluid. These occupy fractures that migrate upward and reseal behind the advancing fluid packet, because the height of fracture that can remain open is controlled by rock strength, as in Fig. 5(a). Hence, the model is essentially one of one-way fluid flow but with the additional factor that at the crack tip $P_f > P_l$ and fluid will tend to be forced out into the wall rock, while at the crack tail the reverse situation obtains, and fluid tends to drain back into the crack. This provides an additional segregation component to contribute to the total amount of quartz in the vein, over and above that precipitated because of the changing temperature of the fluid. A pumping process of this type would primarily deplete the immediate vein walls in quartz, and so it can be ruled out as a major factor in the formation of large veins in Connemara, although it could be an important process in the formation of small veins with a significant segregation component from the walls.

In the case of lower grade rocks that are relatively strong, it is possible to envisage other types of tectonically induced pumping, e.g., between limbs and hinge of a developing fold, that could allow the same fluid to be reused as a transporting medium. However, there is to date no evidence to indicate such a process in the Connemara Schists.

One-Way Fluid Flow

Fluid flow may be strongly channelled into the more permeable horizons, and in addition to fluid released by devolatilization reactions there may be an input of pore fluid driven out of all lithologies as a result of reduction in porosity during syntectonic recrystallization. Hence, although the amounts of fluid apparently required to produce the observed quartz veins in pelites are very large, they may not be wholly impossible for a one-way flow model provided the pelites were the most permeable rocks present. Clearly, the in situ permeability of the pelite layers was very low prior to fracturing, but once a network of spaced cracks had formed, the large increase in permeability that this would cause (Brace, 1980) would permit a drop in fluid pressure within the pelite and hence a gradient that would drive fluid into the fracture network in the pelite from the surrounding rocks.

This is the only mechanism of vein formation for which temperature and pressure differences along the path of fluid flow can provide an adequate driving force for precipitation of quartz. Evidence in support of an important contribution to the Connemara pelite veins by fluids derived from other rock types is provided by the diverse fluid inclusion population present within the vein quartz (Yardley, 1983).

Lateral Segregation

Diffusion through a static pore fluid provides a means of producing segregation veins provided there is a local gradient in chemical potential. However, the field evidence precludes the possibility that this mechanism was important in the generation of the major veins.

Summary

The occurrence of abundant large masses of vein quartz in pelitic schists, with little evidence of a local segregation origin, appears difficult to explain without a contribution from one-way fluid flow. The transient development of convection cells of pore fluid beneath marble or amphibolite cap rocks can also not be ruled out, but the size of such cells is limited by rock strength to a few tens of meters in height. Quartz veins contain a variety of fluid types in inclusions, and it appears likely that they may represent the major pathways for fluid loss, especially during D_3 deformation. It is proposed that initial hydrofracturing to produce veins was a response to reduction of bulk rock permeability resulting from syntectonic recrystallization while fluid continued to be released by reactions. Once the fractures were created, however, they may have raised the permeability of the pelite horizon and hence focused into it flow of fluid from other lithologies. The process was locally probably a cyclic one; individual veins were short-lived. Competency contrasts between pelite and thin siliceous beds may have localised the sites at which veins formed.

Retrograde Veins of Knockaunbaun

One of the most consistent features of the prograde quartz veins is the sparsity of associated metasomatism. However, a distinctive set of quartz veins in Knockaunbaun townland (Fig. 2) are characterized by extensive metasomatism of their wall rocks. The veins occur within a region of chlorite–muscovite–albite schists, sometimes with biotite, that are products of retrogression of amphibolite facies pelite. Evidence for this includes the occurrence of staurolite schist within 250 m, relic oligoclase cores in albite grains, and a thin (2 m) horizon with distinctive sericitized cordierite poikiloblasts.

Most veins are of quartz, with accessory tourmaline, especially at the margins, in many instances. Quartz-sericitized albite–chlorite veins are also present, though rare, and lack metasomatic wall rocks. The main quartz veins form an en echelon array extending for at least 300 m, with individual veins about 50 m in length and 10–30 cm in width, but with numerous small branches and apophyses (Fig. 3(b)). The wall rocks appear distinctly yellow for a distance of 2–10 cm from the veins, owing to abundant staurolite, and coarse prismatic andalusite is also sometimes present in this inner zone (Fig. 3(b)). The full zonal sequence observed around the veins extends over a distance of 10–30 cm and is summarized in Fig. 6. The essential features are that staurolite occurs with tourmaline, garnet, chlorite, muscovite, minor green biotite, and quartz in the inner wall rocks, and all Fe-Mg minerals are very rich in Fe. Moving out, staurolite vanishes although garnet persists, and muscovite is more abundant. Garnet is richer in Mn in this outer region.

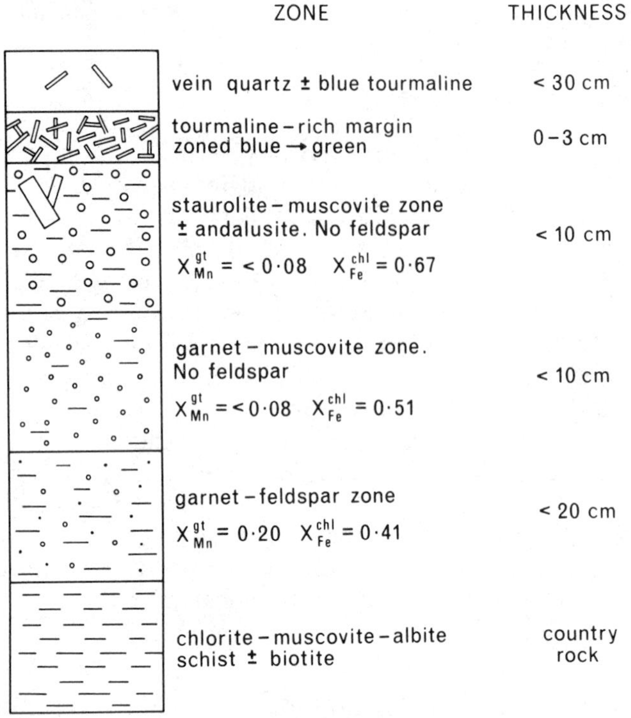

ZONE   THICKNESS

	vein quartz ± blue tourmaline	< 30 cm
	tourmaline – rich margin zoned blue → green	0–3 cm
	staurolite – muscovite zone ± andalusite. No feldspar $X_{Mn}^{gt} = <0·08$ $X_{Fe}^{chl} = 0·67$	< 10 cm
	garnet – muscovite zone. No feldspar $X_{Mn}^{gt} = <0·08$ $X_{Fe}^{chl} = 0·51$	< 10 cm
	garnet – feldspar zone $X_{Mn}^{gt} = 0·20$ $X_{Fe}^{chl} = 0·41$	< 20 cm
	chlorite – muscovite – albite schist ± biotite	country rock

Fig. 6. Schematic representation of the mineral zoning adjacent to quartz veins at Knockaunbaun. Representative mineral compositions are indicated.

Still further from the vein, garnet disappears and albite appears for the first time to give the normal country rock assemblage. An obvious interpretation might be that garnet and staurolite have persisted from the earlier regional metamorphism in the vein walls only, as a consequence of external control of the local fluid composition; however, their compositions are distinctly more Fe rich than most regional metamorphic grains (Yardley *et al.*, 1980), and garnets show only weak internal zoning but vary from almandine close to the vein to spessartine rich only about 15 cm further out. Furthermore, inclusion trains suggest that these porphyroblasts grew later than the regional grains, after the D_3 deformation. Hence, it is concluded that the garnets and staurolite grew during the veining and are a consequence of the metasomatic, and perhaps also thermal, effects of the fluid that produced the quartz veins.

It seems inevitable that a fluid that could cause staurolite, garnet, and andalusite to grow in the vein walls must have been derived itself from pelitic schist, as a result of continuing metamorphism and dehydration at depth or along strike, and this is consistent with what is known of the thermal history of the region. The precise composition of a chloride solution in equilibrium with pelitic assemblages varies as a function of temperature, with the hotter fluid being richer in KCl, $FeCl_2$, and HCl and depleted in NaCl compared to the equilibrium composition at lower temperatures (Eugster and Gunter, 1981). Of these differences, the amount of K-Na exchange when fluid passes from hot pelite will be much greater than for the other possible metasomatic effects, because the concentrations of K and Na in solution are much greater for almost all possible chloride fluids derived from pelites. Only very small amounts of reaction are needed for the HCl content of the migrating fluid to reequilibrate, and so H-metasomatism is only possible where fluid–rock ratios are very large.

The changes observed in the field appear to accord extremely well with the predicted effects of passing fluid from hotter pelite to cooler. The zone of K-metasomatism (albite replaced by muscovite) is the widest, while andalusite growth, reflecting replacement of alkalies by H, is the most restricted. The inner wall rocks show strong Fe enrichment. Table 1 presents the results of calculation of the dominant alkali metal concentration in solution, assuming a 2 m chloride fluid, together with H^+ and HCl concentrations. Fluid compositions have been calculated from the data and method of Frantz et al. (1981) for fluid at 650°C and 2 kb in equilibrium with muscovite–albite–quartz–kyanite, to represent the source composition, and for fluid in the same assemblage at 500°C to represent the final composition after equilibration. There are considerable uncertainties in this data set, and so too much reliance cannot be placed on these calculations; however, for the particular subsystem used in the calculations presented here, the experimental constraints on the data set are in fact quite good. The main limitation of the approach of Frantz et al. is that it combines mass action and mass balance equations, but makes the assumption that activity coefficients are equal to unity. This is a serious error for the charged species, and while activity coefficients will cancel in the case of mass action equations, it is clearly erroneous to use this assumption in the mass balance equations. An estimate of the errors introduced through this assumption has been made using approximate activity coefficients for charged species calculated for vapor-saturated liquid water of equivalent density to the metamorphic fluids using the data of Henley et al. (1984). These calculations suggest that, while the values for the molalites of particular species calculated by the method of Frantz et al. (1981) will be incorrect, the figures for total Na and total K are not greatly in error. This is in accordance with the findings of Vidale (1983), who ob-

Table 1. Calculated molal concentrations in a 2 m chloride fluid in equilibrium with quartz + andalusite + muscovite + albite

	Case A (fluid sink) $T = 723$ K, $P = 2$ kb	Case B (fluid source) $T = 923$ K, $P = 2$ kb
$NaCl + Na^+$	1.96	1.38
$KCl + K^+$	0.04	0.60
$HCl + H^+$	1×10^{-3}	0.02
pH	4.1	4.5

tained experimental fluid composition in terms of total alkali metal contents that agreed with predictions from the method of Frantz *et al.* (1981). In recognition of the uncertainties, details of the speciation are not presented in Table 1.

The differences between the two calculated fluid compositions yields the amount of material that may be added to or removed from the wall rock as the fluid reequilibrates at lower temperatures. The most intensely metaso-matized rocks with over 20% andalusite require fluid–rock ratios of the order of 100 : 1, if the andalusite is the product of H-metasomatism of musco-vite. Further into the wall rocks, however, the amount of fluid required to convert the albite present in the country rocks to muscovite is very much smaller. Fluid–rock ratios decline to about 1 : 1 within 30 cm of the vein edge in most cases.

An obvious problem that remains is why the Knockaunbaun veins should have given rise to such extensive metasomatic effects in their walls, when most prograde veins have very little wall rock alteration associated with them. One possible explanation is that, since the Knockaunbaun veins formed during cooling the fluid pressure in the country rocks had already declined below lithostatic as a result of absorption of pore water in retro-grade reactions. In this case, the fluid pressure in the vein would be higher than that in the country rock and would drive fluid into the vein walls. In contrast, when fractures form during prograde metamorphism the fluid pres-sure in the cracks will be if anything less than that in the country rocks, and fluid will flow out of the walls into the crack, rather than the other way round, thereby preventing fluids passing along the crack from interacting with its walls.

Conclusions

Quartz veins were important, but individually short-lived, pathways for fluid flow through pelitic horizons in the Connemara Schists. They developed most extensively when ductile deformation occurred at the same time that fluid was being released by reactions. At other times in the metamorphic history, extensive fluid loss may have occurred by pervasive flow along microcracks throughout the bulk rock. The reason why veins developed at some times but not others is interpreted as resulting from a reduction in permeability due to syntectonic recrystallization accompanying folding, which caused fluid pressure to rise as volatiles were given off by reaction until failure occurred. Changes in rock permeability are likely to have a far greater influence on whether or not veins form than fluctuations in the rate of volatile release, because permeability can change over a very large range.

There is some uncertainty in estimating the amount of fluid needed to precipitate the quartz veins observed, but the quantities involved are large and over some groups of outcrops amount to an overall fluid–rock ratio in excess of 1 : 1. This may indicate that local segregation and recirculation processes might have operated, but large-scale convective circulation is not possible in a system where P_f is substantially greater than hydrostatic and so one-way flow was probably important. It is inferred that when spaced fractures did form in the pelite layers, they raised the permeability of the layer as a whole so that it became a focus for flow of fluid derived from adjacent rock units.

Acknowledgments

This work was initiated at the School of Environmental Sciences, University of East Anglia, and largely completed at the Institut für Mineralogie und Petrographie, E.T.H. Zurich. I am indebted to N.E.R.C., The Royal Society, and the University of East Anglia for support. The ideas here have benefited greatly from discussions with Alan Thompson, John Ridley, Ernie Perkins, and Larryn Diamond; the mistakes are my own.

References

Barber, J.P., and Yardley, B.W.D. (1985) Conditions of high grade metamorphism in the Dalradian of Connemara, Ireland. *J. Geol. Soc. London* **142**, 87–96.

Brace, W.F. (1968) The mechanical effects of pore pressure on the fracturing of rocks, in *Research in Tectonics,* edited by A.J. Baer and D.K. Norris. *Geological Survey of Canada Paper* **68-52**, 113–124.

Brace, W.F. (1980) Permeability of crystalline and argillaceous rocks. *Int. J. Rock Mech. Min. Sci.* **17**, 241–251.

Brace, W.F., and Orange, A.S. (1968) Electrical resistivity changes in saturated rocks during fracture and frictional sliding. *J. Geophys. Res.* **73**, 5407–5420.

Brace, W.F., Walsh, J.G., and Frangos, W.T. (1968) Permeability of granite under high pressure. *J. Geophys. Res.* **73**, 2225–2236.

Bruton, C.J., and Helgeson, H.C. (1983) Calculation of the chemical and thermodynamic consequences of differences between fluid and geostatic pressure in hydrothermal systems. *Amer. J. Sci.* **283-A**, 540–588.

Cobbold, P.R., Cosgrove, J.W., and Summers, J.M. (1971). Development of internal structures in deformed anisotropic rocks. *Tectonophysics* **12**, 25–53.

Etheridge, M.A., Wall, V.J., and Vernon, R.H. (1983) The role of the fluid phase during regional deformation and metamorphism. *J. Metam. Geol.* **1**, 205–226.

Etheridge, M.A., Wall, V.J., and Cox, S.F. (1984) High fluid pressure during regional metamorphism and deformation: Implications for mass transport and deformation mechanisms. *J. Geophys. Res.* **89**, 4344–4358.

Eugster, H.P., and Gunter, W.D. (1981) The compositions of supercritical metamorphic solutions. *Bull. Soc. Fr. Min. Cryst.* **104**, 817–826.

Ferry, J.M. (1976) P, T, f_{CO_2} and f_{H_2O} during metamorphism of calcareous sediments in the Waterville-Vassalboro area, south-central Maine. *Contrib. Mineral. Petrol.* **57**, 119–143.

Ferry, J.M. (1980) A case study of the amount and distribution of heat and fluid during metamorphism. *Contrib. Mineral. Petrol.* **71,** 373–385.

Ferry, J.M. (1983) Regional metamorphism of the Vassalboro Formation, south-central Maine, U.S.A.: A case study of the role of fluid in metamorphic petrogenesis. *J. Geol. Soc. London* **140,** 551–576.

Fisher, G.W. (1978) Rate laws in metamorphism. *Geochim. Cosmochim. Acta* **42,** 1035–1050.

Frantz, J.D., Popp, R.K., and Boctor, N.Z. (1981) Mineral-solution equilibria. V. Solubilities of rock-forming minerals in supercritical fluids. *Geochim. Cosmochim. Acta* **45,** 69–77.

Fyfe, W.S., Price, N.J., and Thompson, A.B. (1978) *Fluids in the Earth's Crust.* Elsevier, Amsterdam.

Graham, C.M., Greig, K.M., Sheppard, S.M.F., and Turi, B. (1983) Genesis and mobility of the H_2O-CO_2 fluid phase during regional greenschist and epidote amphibolite facies metamorphism: A petrological and stable isotope study in the Scottish Dalradian. *J. Geol. Soc. London* **140,** 577–600.

Hambray, M.J., and Milnes, A.G. (1975) Boudinage in glacier ice—some examples. *J. Glaciol.* **14,** 383–393.

Henley, R.W., Truesdell, A.H., and Barton, P.B., Jr. (1984) Fluid-mineral equilibria in hydrothermal systems. *Reviews in Economic Geology 1.* Soc. Econ. Geol. El Paso.

Leake, B.E. (1969) The origin of the Connemara migmatites of the Cashel district, Connemara, Ireland. *Quart. J. Geol. Soc. London* **125,** 219–276.

Leake, B.E. (1981) 1:63,360 Geological map of Connemara, Ireland. University of Glasgow.

Leake, B.E., Tanner, P.W.G., and Senior, A. (1975) The composition and origin of the Connemara dolomitic marbles and ophicalcites, Ireland. *J. Petrol.* **16,** 237–270.

Norton, D., and Knight, J. (1977) Transport phenomena in hydrothermal systems: Cooling plutons. *Amer. J. Sci.* **277,** 937–981.

Platt, J.P., and Vissers, R.L.M. (1980) Extensional structures in anisotropic rocks. *J. Struct. Geol.* **2,** 397–410.

Rumble, D., III, and Spear, F.S. (1983) Oxygen-isotope equilibration and permeability enhancement during regional metamorphism. *J. Geol. Soc. London* **140,** 619–628.

Rumble, D., III, Ferry, J.M., Hoering, T.C., and Boucot, A.J. (1982) Fluid flow during metamorphism at the Beaver Brook fossil locality, New Hampshire. *Amer. J. Sci.* **282,** 886–919.

Rye, R.O., Schuiling, R.D., Rye, D.M., and Jansen, J.B.H. (1976) Carbon, hydrogen and oxygen isotopic studies of the regional metamorphic complex at Naxos, Greece. *Geochim. Cosmichim. Acta* **40,** 1031–1049.

Secor, D. (1968) Mechanics of natural extension fracturing at depth in the earth's crust, in *Research in Tectonics,* edited by A.J. Baer and D.K. Norris. *Geological Survey of Canada Paper* **68-52,** 3–48.

Thompson, A.B. (1983) Fluid-absent metamorphism. *J. Geol. Soc. London* **140,** 533–548.

Tracy, R.J., Rye, D.M., Hewitt, D.A., and Schiffries, C.M. (1983) Petrologic and stable isotopic studies of fluid-rock interactions, south central Connecticut. I. The role of infiltration in producing reaction assemblages in impure marbles. *Amer. J. Sci.* **283-A,** 589–616.

Vidale, R. (1983) Pore solution compositions in a pelitic system at high temperatures, pressures and salinities. *Amer. J. Sci.* **283-A**, 298–313.

Walther, J.V., and Helgeson, H.C. (1977) Calculation of the thermodynamic properties of aqueous silica and the solubility of quartz and its polymorphs at high pressures and temperatures. *Amer. J. Sci.* **277**, 1315–1351.

Walther, J.V., and Orville, P.M. (1982) Volatile production and transport in regional metamorphism. *Contrib. Mineral. Petrol.* **79**, 252–257.

Yardley, B.W.D. (1975) On some quartz-plagioclase veins in the Connemara Schists, Ireland. *Geol. Mag.* **112**, 183–190.

Yardley, B.W.D. (1977a) The nature and significance of the mechanism of sillimanite growth in the Connemara Schist, Ireland. *Contrib. Mineral. Petrol.* **65**, 53–58.

Yardley, B.W.D. (1977b) Relationships between the chemical and modal compositions of metapelites from Connemara, Ireland. *Lithos* **10**, 235–242.

Yardley, B.W.D. (1983) Quartz veins and devolatilization during metamorphism. *J. Geol. Soc. London* **140**, 657–663.

Yardley, B.W.D., Leake, B.E., and Farrow, C.M. (1980) The metamorphism of Fe-rich pelites from Connemara, Ireland. *J. Petrol.* **21**, 365–399.

Yardley, B.W.D., Shepherd, T.J., and Barber, J.P. (1983) Fluid inclusion studies of high grade rocks from Connemara, Ireland, in *High Grade Metamorphism, Migmatites and Melting,* edited by M.P. Atherton and C. Gribble, 110–126. Shiva, Orpington.

Chapter 6
Oxygen Isotope Systematics of Quartz–Magnetite Pairs from Precambrian Iron Formations: Evidence for Fluid–Rock Interaction during Diagenesis and Metamorphism

R.T. Gregory

Introduction

The purpose of this paper is to review mineral-pair systematics in light of the types of isotopic effects that can be expected to be recorded by mineral pairs during prograde and retrograde metamorphic events. The oxygen isotopic ratios of minerals in the crust are sensitive to changes in temperature and changes in the isotopic composition of the bulk rock brought about by phase changes induced during metamorphism. In crustal environments, solid-state diffusion rates are slow enough that the presence of a fluid phase is required to promote isotopic exchange on geologic time scales. In rock-buffered systems, mineral pairs will essentially behave as isotopic closed systems, because the fluid composition will be strongly dependent upon local isotopic composition of the rocks. The degree to which fluid–rock interaction is important in a particular metamorphic terrane will be recorded in the stable isotopic systematics first by a failure to satisfy conservation of mass constraints on mineral isotopic composition imposed by protolith bulk composition, and second by the degree and scale over which the isotopic compositions of the minerals are homogenized. Mineral-pair behavior that deviates from closed-system predictions results from the differing exchange rates between minerals and fluids (Gregory and Taylor, in press a, in press b), through the infiltration of fluids whose isotopic composition is externally controlled (Taylor et al., 1963; Rumble et al., 1982; Graham et al., 1983), or combinations of both.

Rock types such as Precambrian iron formations are ideal systems to study because: (1) their modal oxygen is predominantly represented by quartz and an iron oxide phase; (2) the large variations in modal mineralogy result in differences in the bulk $\delta^{18}O$ composition of the primary sedimentary layers; (3) the same mineral assemblage is stable over a wide range of temperatures and pressures; and (4) the silica and iron oxide precursors are presumed to have precipitated from seawater or other surface waters suggesting that the initial ^{18}O compositions of the phases were probably uniform or had restricted compositional variation within single terranes.

Previous workers have concentrated on the secular variation of the ^{18}O composition of Precambrian cherts (e.g., Perry *et al.*, 1978) in order to infer the secular variation of either surface temperature or the oxygen isotopic composition of the oceans. In such studies, the diagenetic and metamorphic events exhibited by the cherts have to be discounted or shown to have not altered the original isotopic composition of the cherts. In the discussion that follows, the systematics of quartz–magnetite pairs will be used to suggest that most, if not all, of the iron formations analyzed have suffered open-system exchange events during early diagenesis or during later greenschist-to-amphibolite facies metamorphism. Disequilibrium effects similar to those observed in hydrothermally altered rocks (Taylor and Forester, 1979; Gregory and Taylor, 1981; Criss and Taylor, 1983) are typical of the least deformed iron formations, whereas equilibrium between mineral pairs is typical in terranes that have been deformed. In all of the terranes, if the rocks have behaved as closed systems, the observed spread in magnetite and quartz $\delta^{18}O$ appears to be insufficient to be explained by a protolith precipitating out of a low-temperature (<50°C) mass of surface water whose $\delta^{18}O$ composition is similar to that of modern seawater. The data are consistent with requiring a change both in surface temperatures and seawater ^{18}O (e.g., Perry *et al.*, 1978), or alternatively, that the metamorphic events, because of their open-system character, have erased the isotopic record of the surface history of the rocks.

Mineral-Pair Systematics

The exchange systematics of oxygen isotopes in mineral pairs depends upon temperature, reservoir, and kinetic effects. All of these effects have been reviewed recently in Gregory, Criss, and Taylor (in preparation). Mineral-pair data represented in δ versus δ plots can be easily interpreted in terms of mass- and temperature-dependent fractionation effects. Mineral pairs exchanging in closed systems follow trajectories upon heating or cooling that are determined by the bulk ^{18}O composition of the exchanging volume.

The isotopic exchange behavior of mineral pairs can be reduced to a few simple end-member cases in δ–δ diagrams: (1) simple two-phase closed systems (Fig. 1); (2) open systems where the relative exchange rate between the

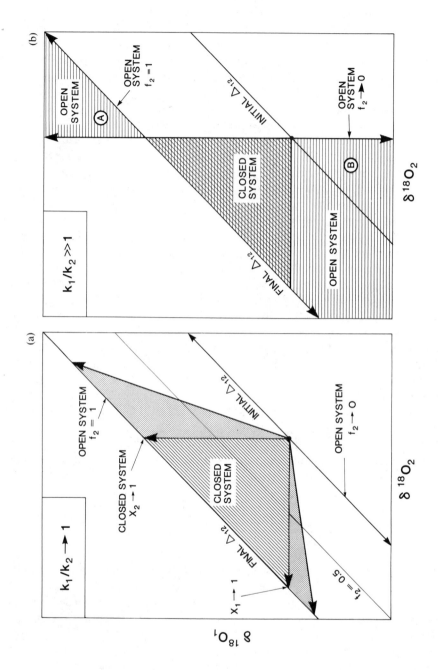

plotted minerals and the external phase is unity (Fig. 1(a)); and (3) open systems where the relative exchange rate between the plotted minerals and the external phase is much greater than one (Fig. 1(b)). In all cases, the area of accessible δ space is constrained by temperature and mass balance.

$$\Delta_{ik} \simeq \frac{A_{ik}}{T^2} + B_{ik} \tag{1a}$$

$$\delta_{sys} = \sum_{i=1}^{n} X_i \delta_i, \tag{1b}$$

where A_{ik} and B_{ik} are constants that reflect the temperature dependence of the fractionation factors derived from isotopic exchange reactions, and where the X_i are the oxygen mole fractions of the individual phases.

In a simple two-phase closed system, the accessible δ space is limited to the northwest quadrant (prograde case $\Delta_{12} < 0$) of a coordinate system centered on the initial point. The boundaries of the fields of δ space accessible in a closed system are determined by vertical and horizontal lines passing through the initial point (these lines to correspond to the mole fractions of the x-axis and y-axis minerals equal to 1, respectively) and a diagonal line forming the hypotenuse of a right triangle. The hypotenuse of the right triangle is approximately an isotherm given by $\Delta_{12} = \delta_1 - \delta_2$. Initially homogeneous mineral pairs found in layers whose modal mineralogy differs (such as mineral pairs that crystallized from a large homogeneous reservoir such as seawater) will upon retrograde or prograde metamorphism lie along line segments colinear with the isotherm marking the temperature of equilibration. The length of the line segment will be proportional to the difference between the temperatures of formation and reequilibration. At the new temperature of isotopic equilibration (final Δ_{12}), the position of the mineral-pair point is determined solely by the modal abundance of each phase, and the difference between initial and final Δ_{12}. Assemblages dominated by mineral 2

◁ *Fig. 1.* Schematic $\delta^{18}O_1$ versus $\delta^{18}O_2$ diagrams illustrating closed-system mineral-pair behavior and open-system behavior for two cases: (a) $k_1/k_2 = 1$; and (b) $k_1/k_2 \gg 1$. For a rock composed only of minerals 1 and 2, and equilibrated at $\Delta_{12}^{initial} \leq \Delta \leq \Delta_{12}^{final}$, all closed systems are required by mass balance considerations to lie within the diagonally ruled right triangle labeled "closed system." The north and west arrows correspond to the trajectories followed by an exchanging mineral pair as the mole fractions of minerals 2 and 1 going to unity, respectively. In the open-system case, when the relative rate constant is unity (a), all partially exchanged mineral pairs will yield measured Δ values that lie between the initial and final equilibrium Δ values. Lines of equal exchange for values of $f_2 = 0, 0.5$, and 1 are shown. The stippled area represents the field of accessible δ space for a system subjected to heterogeneous fluid compositions that are externally controlled. In (b), the relative rate constant is > 1 and the $f_2 = 0$ line becomes vertical. The horizontally ruled area corresponds to the field of accessible δ-space. The mineral pairs that lie within areas labeled A and B will give measured Δ values that give geologically unrealistic temperatures.

will lie at the intersection of a north arrow emanating from the initial point and the new isotherm (marked by $X_2 \rightarrow 1$ in Fig. 1(a)), whereas assemblages dominated by mineral 1 will lie at the intersection of a west arrow and the new isotherm.

Open-system behavior is more complex because once an additional phase is considered, relative exchange rates will determine the geometry of the possible δ space. When describing open systems, it is convenient to define f, the fractional amount of exchange given by:

$$f \equiv \frac{\delta^0 - \delta}{\delta^0 - \delta^{eq}}, \tag{2}$$

where δ^0 refers to the initial mineral δ value and δ^{eq} refers to the final equilibrium value, and $0 < f < 1$. In Fig. 1(a), $f_2 \rightarrow 0$ corresponds to the initial Δ_{12} line and $f_2 = 1$ corresponds to equilibrium at the final Δ_{12}. When the relative exchange rate constant is equal to 1, lines of constant f_2 parallel isotherms, and the area of accessible δ space is determined by the spread in fluid δ values. In the $k_1/k_2 = 1$ case, all mineral pairs yield measured Δ values that give apparent "temperatures" between the initial temperature and the final equilibrium temperature.

When the relative exchange rate (k_1/k_2) is $\geqslant 1$, the $f_2 \rightarrow 0$ line becomes vertical (Fig. 1(b)) and intersects the $f_2 = 1$ line (and all other lines of constant f_2) at the temperature of exchange (final Δ_{12}). Inspection of Fig. 1(b) indicates that for fluids with a markedly different δ value than the initial rock, some measured Δ values will give temperatures that are completely unrealistic (fields labeled A and B, Fig. 1(b)). In natural systems, where temperatures are high, fields A and B (Fig. 1(b)) may correspond to samples that exhibit reversed or abnormal measured Δ values. For example, clino-pyroxene–plagioclase pairs from layered gabbro complexes that have exchanged with circulating groundwaters typically exhibit measured $\Delta_{\text{clinopyroxene–plagioclase}} < 0$ instead of the expected (on thermodynamic grounds) positive-definite values (Taylor and Forester, 1979; Gregory and Taylor, 1981).

The simple open- and closed-system examples discussed above are easily interpreted on δ–δ diagrams because temperature, mass balance, and kinetic effects can be represented simultaneously. A more detailed description of mineral-pair behavior that forms the foundation for discussion of the Precambrian siliceous iron formation $\delta^{18}O$ data is presented below.

Closed Systems

In a simple two-phase closed system, manipulation of the material balance equation (eq. (1b)) shows that exchange trajectory of a mineral pair is given by:

$$\delta^{18}O_1 = -\frac{X_2}{X_1} \delta^{18}O_2 + \frac{\delta_{\text{sys}}}{X_1}. \tag{3}$$

In $\delta_1 - \delta_2$ space, the slopes of all exchange trajectories in such a simple two-phase closed system are negative. Where more than two phases are involved, closed-system mineral-pair exchange trajectories are still linear (Gregory and Taylor, in press a) and are functions of the mole fractions of the phases, and the temperature coefficients (assuming a $1/T^2$ dependence) given in Eq. (1a):

$$\delta_1 = -M\delta_2 + N, \tag{4}$$

where

$$M \equiv \frac{X_2 + \sum_{j=3}^{n} X_j\left(1 + \frac{A_{2j}}{A_{12}}\right)}{X_1 - \sum_{j=3}^{n} X_j\left(\frac{A_{2j}}{A_{12}}\right)}$$

$$N \equiv \frac{\delta_{\mathrm{sys}} - \sum_{j=3}^{n} X_j\left(B_{12}\frac{A_{2j}}{A_{12}} - B_{2j}\right)}{X_1 - \sum_{j=3}^{n} X_j\left(\frac{A_{2j}}{A_{12}}\right)}.$$

In multiphase closed systems (Fig. 2), if the chosen mineral pair is not the pair with the largest fractionation, then some exchange trajectories can have positive slopes (Eq. (4)). A vector for each mineral present in the assemblage can be constructed which records the direction that the plotted mineral pair will move in δ space when the temperature is lowered or raised. The direction of these $X_i = 1$ vectors can be calculated from:

$$\underset{X_j \to 1}{\mathrm{Limit}}\ M = -\frac{1 + A_{2j}/A_{12}}{A_{2j}/A_{12}}, \tag{5}$$

where the A factors refer to the $1/T^2$ fractionation factor coefficients between the additional phase and two plotted phases (Gregory and Taylor, in press a). By utilizing the $X \to 1$ vectors in mineral-pair diagrams, the direction of mineral-pair $\delta^{18}O$ change can be qualitatively estimated from the modal mineralogy. The effect of additional phases on the accessible δ space is to transform the right triangular-shaped space into an oblique triangular-shaped space (Fig. 2). By choosing the minerals with the largest Δ (such as quartz and magnetite in iron formation), we guarantee that all closed-system mineral pairs will lie within the northwest and southeast quadrants. In a coordinate system chosen in this manner, all other $X \to 1$ vectors will lie within the right triangular-shaped space, and the effect of these additional phases will tend to restrict the actual accessible δ space to some subset of the northwest and southeast quadrants (Fig. 2).

In a simple two-phase system such as magnetite–quartz (Fig. 3), the following apply: (1) the slope of the $\delta^{18}O$ exchange trajectory will be determined by the modal abundance of the phases; (2) all such exchange trajectories will have negative slopes equal to $-X_{\mathrm{qtz}}/X_{\mathrm{mgt}}$ where the X's refer to the

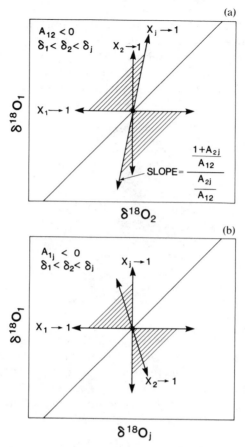

Fig. 2. δ_1 versus δ_2 diagram for an *n*-phase closed system. The right triangular-shaped space that is accessible for all simple two-phase systems has been transformed into an oblique triangle. In this example, the phase represented by X_i is enriched in ^{18}O relative to minerals 1 and 2. Depending on the modal mineralogy of the rock mass, some exchange trajectories in this system (*i*-rich assemblages) could have positive slopes (a). If the $\delta^{18}O$ value of the first phase is plotted against that of the *i*th phase (b), then the closed-system space would revert back to the northwest and southeast quadrants, and the $X_2 = 1$ vector would then point in a northeasterly direction. The magnetite–quartz pair for siliceous iron formations is analogous to the case represented in (b).

mole fractions of oxygen for the particular mineral; and (3) the field of all possible δ values lies within the northwest and southeast quadrants of a coordinate system centered on the initial point.

Open Systems

During metamorphism, because solid-state diffusion rates are sluggish, mineral-pair arrays that lie on isothermal segments that are too short for the

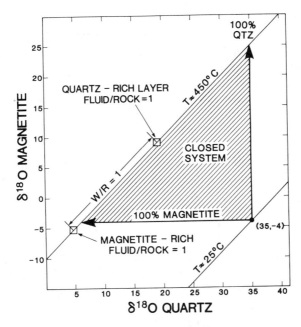

Fig. 3. $\delta^{18}O$ magnetite versus $\delta^{18}O$ quartz diagram illustrating closed-system behavior for mineral pairs precipitated from a seawater reservoir at low temperatures (35, −4). The prograde field is shown for an initial point that lies along the 25°C isotherm. Emanating from the initial point (35, −4) are the 100% magnetite and quartz vectors, the west- and north-pointing vectors, respectively, that represent the legs of a right triangle that bounds the field of accessible δ space for prograde closed-system diagenesis/metamorphism. The intersection of the vectors with the $T \simeq 450°C$ isotherm defines the mineral $\delta^{18}O$ heterogeneity that would be observed if all mineral pairs (in separate sedimentary layers, each of which with a different modal composition) behaved as closed systems. Note that along the isotherm, the mole fraction of quartz will increase monotonically as $\delta^{18}O$ increases. Also shown is the spread along the 450°C isotherm for a system that experiences a hydrothermal exchange event with a $\delta^{18}O = 0$ fluid phase, $w/r = 1$ (oxygen units). As discussed in the text, the effect of the external phase is to limit the observed spread in mineral-pair $\delta^{18}O$ values. The figure was constructed using the fractionation factors of Becker (1971).

inferred retrograde or prograde temperature change indicate the presence of a pervasive metamorphic fluid. Under crustal conditions, complete homogenization of mineral-pair δ values to a constant Δ (a single point in a mineral-pair plot) requires the presence of a large reservoir of infiltrating fluid (Taylor *et al.*, 1963).

When isotopic exchange between isolated rock masses such as permeable zones separated by impermeable zones and some externally ^{18}O buffered fluid occurs, the spread in δ values for any phase along the isotherm is given by:

$$\delta_{max} - \delta_{min} = [1/(1 + w/r)] \, |\Delta^f - \Delta^i|. \tag{6}$$

An infinite w/r results in uniform mineral-pair $\delta^{18}O$ values, whereas in several *independent* simple closed systems (consisting of two phases present in different modal proportions), changing the temperature of equilibration will result in the maximum spread of δ values along the new isotherm (Fig. 3). By replacing w with the amount of oxygen residing in solid phases other than the analyzed mineral pair, it becomes apparent that the greater the number of phases, the more difficult it will be to develop any heterogeneity in individual mineral δ values. Likewise, under typical metamorphic conditions, it will be difficult to impart any dramatic shift in the δ value of the minerals resulting from processes such as Rayleigh distillation when the phase that is removed from the system is but a small part of the total rock oxygen budget (for a review of this problem, see Rumble, 1982).

The calculations above assumed that equilibrium is maintained between all of the phases during any heating or cooling event within the rock. However, once a fluid phase or any additional phase is considered, relative exchange rates must be considered. The behavior of a mineral that is exchanging with a fluid phase of constant composition and at fixed temperature can be described by the equation:

$$\ln(1 - f_j) = -k_j t = \left(\frac{w}{r}\right)_{\text{effective}}, \tag{7}$$

where f_j is defined as the fractional amount of exchange of jth phase, k_j is the rate constant, t is time, and w/r is the effective fluid–mineral ratio (oxygen units) for a system where the external phase is a fluid.

Because the time and exchange rate parameters are difficult to assess for geologic environments, it is convenient to eliminate time from Eq. (7) by comparing two minerals to come up with the expression:

$$1 - f_1 = (1 - f_2)^{K_{12}}, \tag{8}$$

where K_{12} is the relative rate constant k_1/k_2. Eq. (8) when combined with the definition of f (Eq. 2)) yields an expression that is the trajectory for phases 1 and 2 exchanging at different rates with an additional phase such as an externally controlled fluid phase held at a constant δ composition (for complete details see Gregory *et al.*, in preparation).

In real geologic environments, the isotopic composition of the fluid is likely to change as a function of position depending upon the temperature distribution and the $\delta^{18}O$ composition of all of the rocks in the terrane. In order to consider the effect of heterogeneous fluid $\delta^{18}O$ values on the mineral pairs of a particular geologic environment, families of exchange trajectories can be calculated by varying the isotopic composition of the fluid over a wide range of δ values (Fig. 4). Contours of constant f (that connect points of equal fluid/rock on different exchange curves) will be represented as straight lines on mineral-pair diagrams (Fig. 4) whose slope is given by:

$$\frac{1 - (1 - f_2)^{K_{12}}}{f_2}. \tag{9}$$

Because f is always positive and <1, all lines of constant f will have positive slopes in mineral-pair diagrams. These lines of constant f are equivalent to contours of equal amount of exchange or reaction. They also can be thought of as contours of equal fluid/rock or time if relative exchange rates or fluid flux rates are comparable between the individual samples over the area under consideration. If K_{12} is large (>10), the lines of constant f will have slopes that approach $1/f$ and will intersect exchange isotherm at a single point where $\delta_2 = \delta_2^0$. When $K_{12} \gg 10$, the position of the mineral-pair array on a δ–δ plot is clearly independent of δ_1^0, the more rapidly exchanging mineral. In localities where a mineral has only partially exchanged while the coexisting phases have reequilibrated, mineral-pair diagrams involving the partially exchanged phase will reveal positive-sloped data arrays that deviate from slope ~1 behavior. In a rock dominated by two phases, such behavior is clearly indicative of an infiltration event.

For minerals with relative rate constants approaching 1, all partially exchanged mineral pairs will lie along pseudoisotherms (slope ~1 lines) in mineral-pair diagrams (Eq. (9)). Examination of Figs. 1 and 3 indicates that such pseudoisotherms are underestimates of temperature during prograde metamorphism and overestimates during retrograde metamorphism. The potential for pseudoisotherms suggests that all isotopic temperatures should be evaluated in the context of other tests for equilibrium such as textural equilibrium, the homogeneity of mineral chemistry, and concordancy of isotopic ratios determined on mineral triplets (Taylor, 1968; Deines, 1977).

Measurements of $\delta^{18}O$ on coexisting minerals can be used to address problems regarding the degree of equilibration, temperatures of equilibration, and whether the rock mass has behaved as a closed or open system. In order to make maximum use of ^{18}O data, it is *critically important* to have modal information on hand specimen scales and lithologic distributions on outcrop and regional scales. With a knowledge of mineral and rock ^{18}O distribution, the systematics of mineral-pair exchange can be used to construct models of ^{18}O evolution that can be tested in the field. In the discussion that follows, the magnetite–quartz $\delta^{18}O$ distribution in Precambrian iron formations will be examined in light of the systematics presented above.

Precambrian Iron Formation

The origin of the Precambrian banded iron formations is important to our understanding of the history of the oceans and the atmosphere (e.g., Cloud, 1968). The occurrence of chert and magnetite in these types of regularly layered sediments provided potential prospects for paleoclimatological research. In principle, the Precambrian iron formations because of their multiphase assemblages have greater paleoclimatological potential than many Phanerozoic sedimentary rocks where only a single phase, such as carbonate or quartz, is typically analyzed. The correlation between $\delta^{18}O$ of chert and

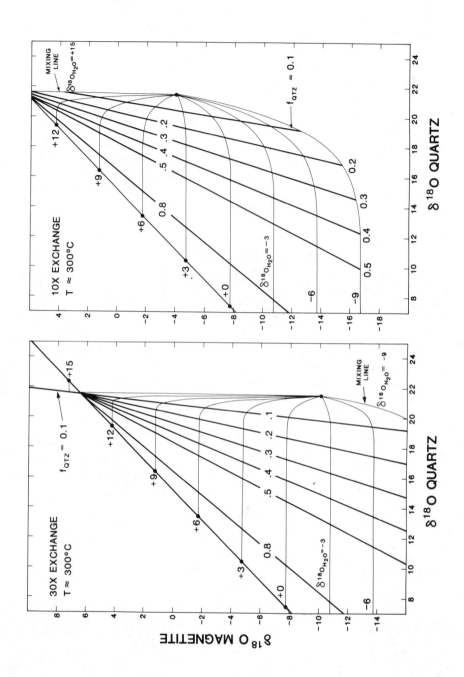

sample age has given rise to conflicting interpretations that can be summarized by grouping the hypotheses into two camps: (1) the $\delta^{18}O$ of the oceans has been variable back through time (e.g., Perry *et al.*, 1978); and (2) the $\delta^{18}O$ of the oceans has remained relatively constant, and thus the secular change in the chert data reflects changes in formation temperature (Knauth and Epstein, 1976). The trade-off between surface temperature and water $\delta^{18}O$ composition is central to all stable isotope studies of paleoclimatology. Much of the discussion of the Precambrian chert data has been based upon quartz–magnetite data (see Becker and Clayton, 1976; Perry *et al.*, 1978; Perry and Ahmad, 1983). A critical issue in the application of Precambrian chert data to the discussion of the surface history of the Earth is whether these systems have behaved as open or closed systems during their post-depositional history. Clearly, if any of the data sets fail the closed-system test, the applicability of the particular terrane to the discussion of the ^{18}O history of the oceans or the surface temperature is called into question.

A summary of magnetite–quartz data from Precambrian cherts is shown in Figs. 5 and 6. There appear to be two major classes of data: (1) mineral pairs from Brazil, West Greenland, Russia, and Minnesota that lie along slope ~1 arrays and whose $\Delta_{qtz-mgt}$ values are suggestive of equilibration under greenschist-to-amphibolite conditions, and (2) mineral pairs from areas such as Australia, Great Lakes region of North America, and South Africa that lie along steep (nearly vertical) positively sloped arrays indicative of disequilibrium exchange under *open-system* conditions. Mineral pairs that map out the slope ~1 arrays typically come from either deformed terranes that exhibit either slaty cleavage or schistosity, or from contact metamorphic terranes that have clearly been recrystallized (Perry and Bonnichsen, 1966). Mineral pairs that lie along the open system disequilibrium arrays come from low-grade undeformed terranes such as the Hamersley basin that have only suffered a burial metamorphic event (Smith *et al.*, 1982). The undeformed Hamersley group rocks that preserve oxygen isotope disequilib-

---

◁ *Fig. 4.* Calculation of open-system exchange trajectories for the $\delta^{18}O$ in magnetite and quartz for fluid $-9 \leq \delta^{18}O \leq 15$ and two different initial rock $\delta^{18}O$ compositions. The straight lines superimposed on the exchange curves are the lines of constant f_{qtz} (0.1, 0.2, 0.3, 0.4, 0.5, and 0.8). The $\approx300°C$ isotherm is the same as the $f = 1$ line that corresponds to isotopic equilibrium. The initial points (21.5, -4) and (21.5, -10) were chosen arbitrarily. The initial point (21.5, -10) is shown with its associated lines of constant f_{qtz} for relative rate (constant 30×) that incorporates the average model abundance in iron formation of quartz to magnetite (3 : 1) into the relative rate constant assumed to be equal to 10. Note that the position of the lines of constant f are independent of the initial magnetite $\delta^{18}O$ value. The calculations indicate that in systems where fluid $\delta^{18}O$ values have been heterogeneous, samples that have suffered similar amounts of exchange lie along steep positive-sloped arrays in mineral-pair diagrams (see Gregory *et al.*, in preparation). For large values of the relative rate constant, it is possible to use the intersections of the data array with alteration isotherm to infer the initial $\delta^{18}O$ value of the more slowly exchanging mineral.

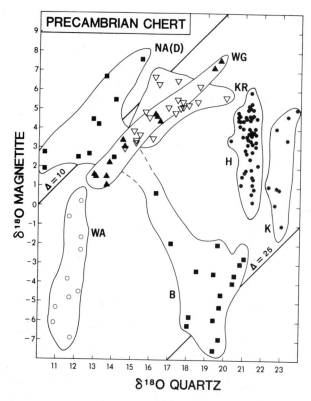

Fig. 5. Precambrian chert data for $\delta^{18}O$ of magnetite and quartz is shown for cherts from West Greenland, WG (Perry *et al.,* 1978); Krivoj Rog, Russia, KR (Perry and Ahmad, 1981); Hamersley basin, Western Australia, H (Becker, 1971; Becker and Clayton, 1976); metachert in the contact aureole of the Duluth gabbro, Minnesota, North America, NA(D) (Perry and Bonnichsen, 1966); other North American cherts from the Great Lakes region, B (James and Clayton, 1962; Perry *et al.,* 1973); Weld Range, Western Australia, WA (Perry and Ahmad, 1983); and Kuruman Formation, South Africa, K (Perry and Ahmad, 1983). These data indicate that under some conditions quartz readily exchanges ^{18}O with its surroundings, particularly in rocks that have been deformed (KR and WG) or have been pervasively recrystallized (NA(D)). In low-grade, relatively undeformed areas (WA, B, H, and K), the data lie along steep disequilibrium arrays that are characteristic of hydrothermally exchanged rocks. The positions of the disequilibrium arrays indicate that different initial ^{18}O compositions are required for the preburial metamorphic event cherts. However, because of the uncertainty resulting from open-system conditions, the initial isotopic composition of the cherts can only be estimated using model dependent calculations.

rium are in contrast to the Brazilian iron formations (Hoefs *et al.*, 1982) that have developed three foliation planes all of which consistently exhibit measurably different average $\Delta_{qtz-mgt}$ values (Fig. 6). It appears that oxygen isotopic exchange in quartz is greatly enhanced by the deformation events and thus deformed terranes commonly approach isotopic equilibrium.

Equilibrium Arrays

In an ^{18}O study of coexisting magnetite and quartz taken from amphibolite facies iron formation from West Greenland, Perry *et al.* (1978) conclude that the quartz–magnetite pairs equilibrated under closed-system conditions. Projection back along the individual West Greenland sample closed-system exchange trajectories using the published modes of Perry *et al.* (1978) possibly suggests that some samples (e.g., $X_{mgt} = 0.52$, 0.53, and 0.09) originally may have had uniform $\delta^{18}O$ values of quartz and magnetite approximately equal to 20 and -5, respectively; $\Delta_{qtz-mg} \approx 25$. This result is shown graphically in Fig. 7(a). If the intersection of $X_{mgt} = 0.53$ and 0.09 vectors (the most magnetite-rich and quartz-rich layers) represents the true, preamphibolite grade initial $\delta^{18}O$ reservoir of the system, then the system, at some time in its history, was homogenized at a temperature of $\sim 100°C$. If the ^{18}O reservoir was recording a diagenetic event and not a surface event, as Perry *et al.* (1978) suggest, then all paleoclimatic significance of the data was lost at this point. The $\delta^{18}O$ composition of the fluid necessary to account for such a homogenization event during diagenesis is within a few per mil of present-day seawater (Fig. 8) and a reasonable $\delta^{18}O$ composition for modern-day groundwaters (e.g., see Taylor, 1974).

An unresolved question is whether the dynamically metamorphosed samples, whose minerals appear to exhibit oxygen isotopic equilibrium, initially passed through an open-system burial history analogous to the terranes such as the Biwabik or the Hamersley. Because the individual West Greenland sample exchange trajectories intersect at more than one point, and because there is no monotonic progression of X_{mgt} values along the data array, the individual layers have not all behaved as closed systems. This open system deviation from the monotonic progression of modal magnetite along the data array (required by mass balance for an initially uniform closed system (see Figs. 1 and 3)) is suggestive of a more complex history involving equilibration on different scales and closure of different layer sets at different temperatures. A disequilibrium array such as the stippled array shown in Fig. 7(a), is capable upon equilibration of generating the observed equilibrium array. Support for this interpretation comes from the Biwabik iron formation (Perry and Ahmad, 1983), which has been contact metamorphosed by the Duluth gabbro (Fig. 7(b)). The Biwabik data demonstrate that precontact metamorphic disequilibrium arrays transform into equilibrium arrays when sufficient thermal energy is applied to the system. Inspection of Fig. (7) indicates that the iron formations that display the nonequilibrium quartz–magnetite ^{18}O

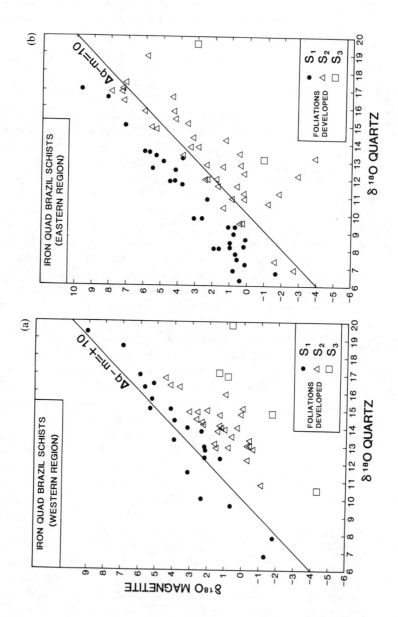

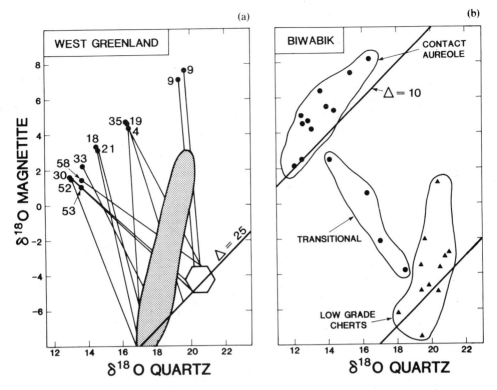

Fig. 7. (a) The modal magnetite oxygen contents of the analyzed mineral pairs are shown at the intersections of mineral-pair exchange trajectories with $T \simeq 100°C$ ($\Delta = 25$) isotherm, for the West Greenland magnetite–quartz $\delta^{18}O$ data of Perry *et al.* (1978). Note that there is no monotonic progression of modal magnetite along the equilibrium isotherm. The exchange trajectories calculated from mass balance intersect at many points indicating equilibration on different scales suggestive of a complex history. If the geotherms corresponding to the largest spread of X_{mgt} along the actual data are shown ($X_{mgt} = .53$ and $.09$), then a homogenization event is inferred at $\Delta_{qtz-mgt} = \simeq 25$. Only a few mineral pairs (from the most quartz- and magnetite-rich layers) have δ values consistent with such a closed-system interpretation. The final array would have just as easily come from a disequilibrium array shown by the shaded area. (b) The Biwabik data are shown to illustrate the effect of a high-temperature event on an earlier formed disequilibrium array.

◁ Fig. 6. $\delta^{18}O$ data for magnetite and quartz from the Iron Quadrangle, Brazil data (Hoefs *et al.*, 1982) are divided into lower grade (western region, (a)) and higher grade (eastern region, (b)). In both regions, at least three foliations are developed, all of which have distinctive ^{18}O signatures in mineral-pair diagrams. Because the modal data were not reported for the individual samples, it is not possible to evaluate whether the S_1 and S_2 deformations occurred under closed systems. However, the data are suggestive that the isotopic composition of the fluid was buffered by the rocks during the later retrograde events. If we assume that the spread in mineral-pair $\delta^{18}O$ results from a wide range of modal compositions, then the spread of the Brazilian chert data is comparable to the spread that is calculated for the $w/r = 1$ case shown in Fig. 3, or is comparable with a closed-system spread for an original protolith temperature of 100°C.

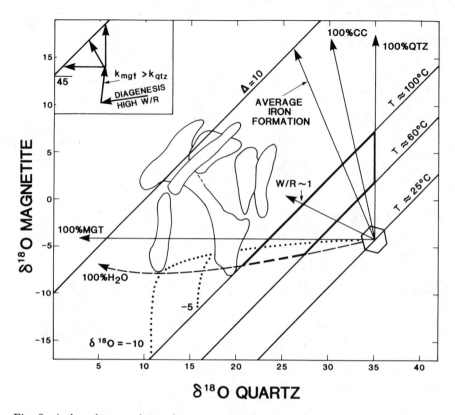

Fig. 8. A thought experiment is constructed for quartz and magnetite precipitated from an ocean similar in temperature and composition to the present-day ocean to infer whether such a protolith could account for the observed chert data (shown by the envelopes corresponding to the data of Fig. 4) given the types of processes that are likely to occur in the crust as a result of diagenesis and burial metamorphism. Arrows showing the direction the plotted mineral pair will evolve during prograde reequilibration are shown for assemblages dominated by the phases: quartz, carbonate, magnetite, water, and for average iron formation (Gole and Klein, 1981). The curve for water and the line for quartz define the maximum spread expected for a low-temperature protolith in equilibrium with an isotopic composition similar to that of modern seawater. The heavy "trapezoidal-shaped" field, bounded by the 60° and 100°C isotherm, schematically represents the temperature window for the opal to quartz transition. If this transition occurs in a chemical environment where the oxide phase is relatively inert, then possible exchange trajectories would be the dotted curves for subsurface fluids having $\delta^{18}O = -10$ and -5. Note that the curve for -10 water can account for the most depleted cherts from the Weld range, Western Australia. The steep positive-sloped data arrays can be explained by processes of the type illustrated in Fig. 4. Note that the entire heterogeneity of the Precambrian chert data can be accounted for by processes that occur in modern sedimentary basins without requiring any major changes to the oxygen isotopic composition of the ancient oceans or in the surface temperature of the earth. Because of the complex metamorphic history of all of the cherts, there is no unique solution to the early isotopic history that is relevant to paleoclimatological studies of the earth.

relationships upon equilibration can easily account for the variation observed in the equilibrated iron formations.

Disequilibrium Arrays

The magnetite–quartz pairs that lie along the steep, disequilibrium arrays cannot be easily modeled. In the igneous examples (Taylor and Forester, 1979; Gregory and Taylor, 1981; Gregory et al. (in preparation)), the initial magmatic $\delta^{18}O$ composition and temperature can easily be estimated, and the exchange event is generally related to the emplacement of the pluton itself. In contrast, the initial $\delta^{18}O$ composition of the mineral pairs in the iron formation case cannot be known with certainty without knowing both the temperatures over which the exchange events occurred and the initial formation temperature of the rocks.

In Fig. 4, families of mixing lines and constant exchange lines are shown for two different initial rock $\delta^{18}O$ compositions and for a temperature of alteration similar to that deduced by Becker and Clayton (1976) and Smith et al. (1982). Examination of Fig. 4 suggests that (1) if the temperature of the exchange event can be deduced, the initial $\delta^{18}O$ values of the quartz can be crudely estimated using the intersection of the data array with the exchange isotherm; and (2) that the positions of the constant f lines are insensitive to the initial $\delta^{18}O$ of the iron oxide phase. Unless there are minor mineral pairs that have equilibrated isotopically or mineralogically (from which a range of fluid compositions and exchange temperature can be calculated), there is no unique way to determine the initial $\delta^{18}O$ value of the slowly exchanging mineral. Even when the initial quartz can be estimated, the $\delta^{18}O$ of the protolith (both minerals) cannot be known *unless the temperature of formation is assumed.*

Discussion

The major conclusion that can be derived from the disequilibrium magnetite–quartz pairs from Australia, North America, and South Africa is that the rocks have behaved as open systems. For a large relative exchange rate constant ($K_{mgt-qtz} \gg 1$), the slope of the disequilibrium array is proportional to $1/f$. Because f is related to the effective water/rock (Eqs. (7) and (8)), the bulk fluid–rock ratio can be estimated from mineral-pair data even though the initial fluid and rock $\delta^{18}O$ are not known. The estimate of the fluid–rock ratio depends upon the choice of the temperature of alteration and hence the relative rate constant. If $f_{qtz} < 0.1$ and, arbitrarily, $K_{mgt-qtz} = 10, 30,$ or 100, the inferred fluid/rock for typical bulk siliceous iron formation (qtz/mgt $\simeq 3$) is 0.2, 0.4, or 1.4 (in weight units), respectively. The fluid/rock only reflects the event that produced the disequilibrium array, with the fluid-flow meter beginning at the time the $\delta^{18}O$ quartz became uniform. Prior to the burial metamorphic event the initial $\delta^{18}O$ values of the quartz in the terranes *were different*. The spread in initial $\delta^{18}O$ composition reflects differing surface water $\delta^{18}O$ values or temperatures at the times of formation of the various iron formations, or alternatively, the heterogeneity reflects some precursor diagenetic event. Most of the evidence regarding the secular ^{18}O variations of the oceans comes from rocks such as the cherts themselves. Arguments regarding the constancy of seawater ^{18}O composition are based mainly upon mass balance arguments concerning the interaction between the oceans and mantle-derived basalts during the accretion of the oceanic crust (Muehlenbachs and Clayton, 1976; Gregory and Taylor, 1981). The potential impact of the second hypothesis is more easily treated through application of the mineral-pair systematics presented above.

In Fig. 8, a thought experiment is constructed to illustrate a plausible and by no means unique sequence of events that would be capable of generating the observed data set. In order to construct the model, the following assumptions have been made: (1) the quartz of the cherts had an amorphous or opaline precursor; (2) groundwaters of the Precambrian were similar in $\delta^{18}O$ composition to those of today; and (3) the ^{18}O compositions of the oceans of the past were similar to those of today. The evolutionary path illustrated in Fig. 8 suggests that early in the history of the iron formation the silica phase transforms from an amorphous precursor phase to an ordered silica polymorph. During this phase transition, the magnetite or other oxide precursor phase may have been relatively inert. There is some evidence from sedimentological and stratigraphic studies of iron formations that the tectonic setting of the depositional basin was some type of shallow epicontinental sea (e.g., see Trendall, 1983). Thus, it is reasonable to suggest that during early diagenesis light meteoric fluids could have had access to these sediments. Mixing lines for 100°C waters exchanging with relatively inert oxide phases and with silica undergoing a phase transition are shown in Fig. 8. Note that these exchange trajectories are similar to the control vector shown for seawater $\delta^{18}O = 0$, and that if the opal–quartz transition occurred at temperatures as high as ~150°C, exchange with waters having $\delta^{18}O$ composition typical of modern-day surface waters ($-10 < \delta^{18}O < 0$), in high water/rock environments, could generate the observed heterogeneity in the samples. In Fig. 8, a second-stage event is inferred to be a burial metamorphic event that probably occurred under lower water/rock conditions as suggested by the spread of the $\delta^{18}O_{mgt}$ data points ($>5‰$) along the disequilibrium arrays, and the limited spread at any locality of the $\delta^{18}O$ quartz data points (<3 per mil). From the scatter of modal composition observed along the slope 1 array (West Greenland case), this second stage event is inferred for the equilibrated cherts that lie along slope ~1 arrays. Finally, the Brazilian samples indicate that significant oxygen exchange occurs for both quartz and magnetite in terranes subjected to multiple metamorphic events that involve penetrative deformation. Considered as a whole, the Precambrian iron formation oxygen isotope data indicate that even the well-preserved iron formations may have suffered multistage oxygen exchange events.

Summary

The oxygen isotope ratios of mineral pairs provide a powerful investigative probe into processes involving fluid–rock interaction because of the sensitivity of the measured $\delta^{18}O$ values to isotopic reservoir, temperature, and kinetic effects. Mineral-pair diagrams provide a means for visualizing reservoir, temperature, and kinetic effects simultaneously. Significantly more information could be inferred from stable isotope studies if quantitative modal information was routinely published with isotopic analyses. Consideration of mineral-pair systematics of simple systems such as iron formations indicates that open system processes involving the redistribution of stable isotopes by fluids is important in the evolution of the crust. Currently available data indicate that postdiagenetic time-integrated fluid–rock ratios >0.5 (oxygen units) are not unreasonable for the iron formations and that subsurface fluids whose composition, on average, is comparable to modern-day surface waters (e.g., seawater) can account for the observed isotopic distribution in chert.

From the discussion of the mineral-pair systematics, it is clear that the oxygen data obtained from the Precambrian cherts yield no clear-cut or unique interpretation of the surface conditions responsible for the generation of the original sediments that now are preserved in the iron formations as weakly to intensely metamorphosed rocks. Given the heterogeneity of the preburial metamorphic event cherts, it is tempting to interpret the heterogeneity as representing a true secular variation in either the surface temperature or the ^{18}O composition of seawater. It is also apparent that by calling on plausible subsurface processes, the ^{18}O heterogeneity can be explained through the application of uniformitarian principles without addressing any of the paleoclimatological issues.

Acknowledgments

This paper has benefited from discussions with many people; in particular, H.P. Taylor, Jr., R.E. Criss, C.B. Douthitt, and S. Epstein, who impressed me with the importance of mass balance constraints. E.C. Perry's review improved an earlier version of the manuscript. This work has been supported by grants from the Research Corporation and the Australian Research Grants Scheme.

References

Becker, R.H. (1971) Carbon and oxygen isotope ratios in iron formation and associated rocks from the Hamersley Range of Western Australia and their implications. PhD thesis, University of Chicago, Illinois. 138 pp.

Becker, R.H., Clayton, R.N. (1976) Oxygen isotope study of a Precambrian banded iron formation, Hamersley Range, Western Australia. *Geochim. Cosmochim. Acta* **40,** 1153–1165.

Cloud, P. (1968) Atmospheric and hydrospheric evolution on the primitive Earth. *Science* **16,** 729–736.

Criss, R.E., Taylor, H.P., Jr. (1983) An $^{18}O/^{16}O$ and D/H study of Tertiary hydrothermal systems in the southern half of the Idaho batholith. *Geol. Soc. Amer. Bull.* **94,** 640–663.

Deines, P. (1977) On the oxygen isotope distribution among mineral triplets in igneous rocks. *Geochim. Cosmochim. Acta* **41,** 1709–1730.

Friedman, I., O'Neil, J.R. (1977) Compilation of stable isotope fractionation factors of geochemical interest. U.S. Geological Survey Professional Paper **440-KK.**

Gole, M.J., Klein, C. (1981) Banded iron-formation through much of Precambrian time. *J. Geol.* **89,** 169–183.

Graham, C.M., Greig, K.M., Sheppard, S.M.F., Turi, B. (1983) Genesis and mobility of the H_2O-CO_2 fluid phase during regional greenschist and epidote amphibolite facies metamorphism: A petrological and stable isotope study in the Scottish Dalradian. *J. Geol. Soc. London* **140,** 577–599.

Gregory, R.T., Taylor, H.P., Jr. (1981) An oxygen isotope profile in a section of Cretaceous oceanic crust. Samail ophiolite. Oman: Evidence for $\delta^{18}O$-buffering of the oceans by deep (>5 km) seawater-hydrothermal circulation at mid-ocean ridges. *J. Geophys. Res.* **86,** 2737–2755.

Gregory, R.T., Taylor, H.P., Jr. (In press a) Oxygen isotope evidence for non-equilibrium metasomatic effects in upper mantle assemblages. *Contrib. Mineral. Petrol.*

Gregory, R.T., Taylor, H.P., Jr. (In press b) Possible non-equilibrium oxygen isotope effects in mantle nodules, an alternative to the Kyser-O'Neil-Carmichael $^{18}O/^{16}O$ geothermometer. *Contrib. Mineral. Petrol.*

Hoefs, J., Muller, G., Schuster, A.K. (1982) Polymetamorphic relations in iron ores from the Iron Quadrangle, Brazil: The correlation of oxygen isotope variations with deformation history. *Contrib. Mineral. Petrol.* **79,** 241–251.

James, H.L., Clayton, R.N. (1962) Oxygen isotope fractionation in metamorphosed iron formations of the Lake Superior region and in other iron-rich rocks, in *Petrologic Studies. Buddington volume.* Geol. Soc. Amer. 217–239.

Knauth, L.P., Epstein, S. (1976) Hydrogen and oxygen isotope ratios in nodular and bedded cherts. *Geochim. Cosmochim. Acta* **40,** 1095–1108.

Muehlenbachs, K., Clayton, R.N. (1976) Oxygen isotope composition of the oceanic crust and its bearing on seawater. *J. Geophys. Res.* **81,** 4365–4369.

Perry, E.C., Bonnichsen, B. (1966) Quartz and magnetite: $^{18}O/^{16}O$ fractionation in metamorphosed Biwabik iron formation. *Science* **153,** 528–529.

Perry, E.C., Ahmad, S.N. (1981) Oxygen and carbon isotope geochemistry of the Krivoj Rog iron formation, Ukrainian SSR. *Lithos* **14,** 83–92.

Perry, E.C., Ahmad, S.N. (1983) Oxygen isotope geochemistry of Proterozoic chemical sediments. *Geol. Soc. Amer. Mem.* **161,** 253–264.

Perry, E.C., Tan, F.C., Morey, G.B. (1973) Geology and stable isotope geochemistry of the Biwabik iron formation. Northern Minnesota. *Econ. Geol.* **68,** 1110–1125.

Perry, E.C., Ahmad, S.N., Swulius, T.M. (1978) The oxygen isotope composition of 3,800 M.Y. old metamorphosed chert and iron formation from Isukasia, West Greenland. *J. Geol.* **86,** 223–229.

Rumble, D. (1982) Stable isotope fractionation during metamorphic devolatilization reaction, in *Characterization of Metamorphism through Mineral Equilibria,* edited by J.M. Ferry, pp. 327–354. *Reviews in Mineralogy 10.* Min. Soc. Amer.

Rumble, D., Ferry, J.M., Hoering, T.C., Boucot, A.J. (1982) Fluid flow during metamorphism at the Beaver Brook fossil locality. *Amer. J. Sci.* **282,** 886–919.

Smith, R.E., Perdrix, J.L., Parks, T.C. (1982) Burial metamorphism in the Hamersley basin, Western Australia. *J. Petrol.* **23,** 74–102.

Taylor, H.P., Jr. (1968) The oxygen isotope geochemistry of igneous rocks. *Contrib. Mineral. Petrol.* **19,** 1–71.

Taylor, H.P., Jr. (1974) The application of oxygen and hydrogen isotope studies to problems of hydrothermal alteration and ore deposition. *Econ. Geol.* **69,** 843–883.

Taylor, H.P., Jr., Forester, R.W. (1979) An oxygen and hydrogen isotope study of the Stenerguard intrusion and its country rocks: A description of a 55 m.y. old fossil hydrothermal system. *J. Petrol.* **20,** 355–419.

Taylor, H.P., Jr., Albee, A.L., Epstein, S. (1963) $^{18}O/^{16}O$ ratios of coexisting minerals in three assemblages of kyanite-zone pelitic schist. *J. Geol.* **71,** 513–522.

Trendall, A.F. (1983) The Hamersley basin, in *Iron-formation facts and problems,* edited by A.F. Trendall and R.C. Morris, pp. 69–130. *Developments in Precambrian Geology 6.*

Chapter 7
The Role of Mineral Kinetics in the Development of Metamorphic Microtextures

J. Ridley and A.B. Thompson

Introduction

If the aim of a metamorphic petrologist is to reconstruct as fully as possible pressure–temperature–time (P–T–t) histories of suites of metamorphic rocks, then a consideration of more than the equilibrium aspects of the phase petrology is necessary. Mineral zonation and the preservation of relict minerals as inclusions in porphyroblasts are two expressions of disequilibrium that have been used to give at least qualitative indications of the evolution of pressure and temperature in a rock (e.g., Råheim and Green, 1974; Thompson *et al.*, 1977; Holland and Richardson, 1979; Spear and Selverstone, 1983). Additional information about rock evolution is available from textures and microstructures, both reaction textures, and textures in the sense of the grain size, habit, or distribution of a newly grown mineral (Ehrlich *et al.*, 1972; Kretz, 1966). The purpose of this paper is to discuss how the kinetic processes involved in metamorphic crystallization influence rock and mineral textures, and hence to suggest how rock textures may give information about P–T–t paths of metamorphism.

The physical principles of texture development in metamorphism using comparisons with textures described for metals, taking the minimization of free energy as the basic force for texture development, have been reviewed by Spry (1969) and Vernon (1976). These approaches, however, take little account of the possible effects of specific reaction kinetics. The effect of reaction kinetics on the textures of rock crystallized from a melt and the influence of the magnitude of equilibrium overstepping has been investigated experimentally by Kirkpatrick *et al.* (1979). Loomis (1983) has shown that reaction kinetics may influence the exact composition and zonation patterns of minerals formed in a metamorphic reaction.

This paper considers the theoretical aspects of how kinetic processes may affect textures and how different kinetic factors may be important in different metamorphic regimes or different segments of a $P-T-t$ path. Are any general textures of the whole mineral assemblage, or specific textures of specific minerals, diagnostic of a certain reaction mechanism (cf. Fisher, 1978), and hence of metamorphism with or without a fluid, large or small overstepping of equilibrium boundaries before nucleation, or of different rates of heating?

We consider that the primary driving force for metamorphism is temperature increase—heat transfer by fluids plays a minor role when integrated over the length and time scale of an orogenic episode. Individual reactions in individual layers of rock may take place because of the infiltration of fluids that are out of chemical equilibrium with the rock (see Chapter 3). The fluid in such a case is derived either from neighboring rocks (Graham et al., 1983) or from deeper in the crust (Ferry, 1983). These fluids are, however, generally themselves the result of temperature-induced prograde metamorphic reactions (Walther and Wood, 1984).

Types of Microstructures

A microtexture is a distinguishable part of the rock fabric or microfabric, the fabric being a function of the size, shape, distribution (and composition) of all minerals present (Sander, 1930). A fabric may be either homogeneous or heterogeneous on the scale of observation. Homogeneity in this sense does not imply that every grain of a certain mineral is everywhere the same size and shape. Rather, it implies that on the scale of interest, there are no areas with, for instance, markedly more elongate grains, markedly larger grains, or marked variations in the concentration of one mineral. There are various types of heterogeneous fabrics that are considered important in metamorphic rocks. These include:

domainal structures: volumes within a rock in which some aspect of the fabric is significantly different from elsewhere

porphyroblasts or poikiloblasts: the presence of certain grains with a grain size significantly greater than that of the other minerals in the rock

segregations: a subclass of domainal structures in which some volumes show a mineralogy different from the rest of the rock

The microtextures of any specific rock are influenced by its actual mineralogy and composition and by the kinetics of individual reactions. The kinetics involve an often complex interaction between the processes of nucleation and those of growth and dissolution. Kinetic processes themselves are principally controlled by temperature, the rate of temperature or pressure change, deformation, and the amount of fluid in the rock.

Unfortunately, knowledge of the physical and chemical properties of the various metamorphic minerals is far from complete. When, for instance, does a mineral dissolve and reprecipitate rather than deform through dislocation mechanisms under the influence of imposed stress? Because of these uncertainties, it is felt that most can be achieved by trying to understand, first, why individual minerals or specific assemblages develop a range of textures, and second, why certain textural types appear favored by certain metamorphic situations. The conditions in two different rocks undergoing the same reaction will not be identical: Different components will be required to diffuse to permit product growth. The matrix grain size and general texture before reaction will be different, as will the amount of fluid in the rock. Not all garnet–staurolite schists, for instance, show the same grain-size ratio of the two porphyroblast phases. Sillimanite appears often to grow at the expense of kyanite through a complex reaction cycle involving dissolution and reprecipitation of phases not directly participating in the reaction (Carmichael, 1969). There are, however, cases described where sillimanite directly replaces kyanite (Hollister, 1969). The factors favoring one or the other mechanism—and hence favoring different final textures—are not fully understood. These aspects of texture development are discussed further in the following sections beginning with the processes of nucleation and growth of metamorphic minerals.

Nucleation and the $P-T$ Overstepping of Mineral Equilibrium Boundaries

Nucleation Rate Laws

The nucleation rate law as derived for metallic system is given in Eq. (1), Appendix I,

$$\dot{n} = A_n \exp\left(-\frac{16\pi\gamma^3}{3\Delta G_r^2 \kappa T}\right). \tag{1}$$

Some aspects of the consequences of the form of this equation on metamorphic processes are discussed by McLean (1965, p. 105). He emphasizes the extremely sharp dependence of the nucleation rate ($\dot{n}$) on the grain interfacial energy (γ) and the free energy of reaction (ΔG_r): a much sharper dependence than a normal first-order kinetic rate equation.

In a particular rock ready to undergo a specific reaction, it can be assumed that the interfacial energy (γ) remains constant—the temperature dependence of γ can be ignored over small temperature intervals. On gradual heating across an equilibrium curve it can be seen by rearranging Eq. (1) that the nucleation rate (ignoring pressure and volume effects) will be a function of the amount of temperature overstepping of the reaction, by a relationship (see Appendix I):

$$\dot{n} \ \alpha \ \exp[(T - T_{eq})^2].\qquad(2)$$

A 10% increase in the magnitude of $(T - T_{eq})$ will give an increase in nucleation rate of 6–7 orders of magnitude. What this implies is that whatever the values of the various constants in Eq. (1), on progressive crossing of an equilibrium curve in P–T space, nucleation will always start effectively abruptly after a certain finite overstepping of the curve.

It is worth noting that a similarly strong dependence of the nucleation rate on the amount of overstepping is predicted if a metamorphic reaction is the result of fluid infiltration. In this case, for small departures from equilibrium,

$$\Delta G_r = n(X - X_{eq})\left[RT \ \ln\!\left(\frac{X}{1 - X}\right) + (1 - 2X)W_g \right],\qquad(3)$$

and hence

$$\dot{n} \ \alpha \ \exp[(X - X_{eq})^2].\qquad(4)$$

Nucleation should start abruptly after the infiltrating fluid has changed the local fluid composition by a certain finite amount.

The dependence of $\dot{n}$ on $(T - T_{eq})$ or $(X - X_{eq})$ is probably greater in Eq. (1) than would be the case for real, heterogeneous rocks. Eq. (1) assumes an ideal homogeneous rock in which all potential sites for nucleation have the same free energy both before nucleation and after. In a real rock there will be variations in interfacial energy (γ) as a result of variations in impurity content (McLean, 1965) or of grain boundary orientation (Porter and Easterling, 1981, p. 124). Additionally, in any divariant or multivariant reaction, the value of ΔG will not only be a function of temperature but also of reaction progress. Once a certain amount of nucleation and growth has taken place, the chances of new nucleation will diminish.

Possible Magnitudes of Overstepping in Metamorphism

In progressive regional metamorphism, it is generally considered that, although the magnitude of the overstepping required for nucleation is unknown, it is small enough never to affect significantly an equilibrium treatment of the assemblages (e.g., Guidotti, 1974). The constant pattern of zonal sequences in regional metamorphic terranes suggests that overstepping is never so great that one mineral fails to nucleate in sequence (Fyfe et al., 1958). This may not necessarily be the case in contact metamorphism (Naggar and Atherton, 1970), in the metamorphism of coarse-grained intrusive rocks (Koons et al., in preparation), or in polyphase metamorphism involving earlier granulites.

Approximate magnitudes of the overstepping required are suggested from two sources: (1) substitution of likely values for the various parameters in Eq. (1), and (2) comparing seeded with unseeded high-pressure–temperature petrological experiments (Fig. 1). The important parameters in Eq. (1) are

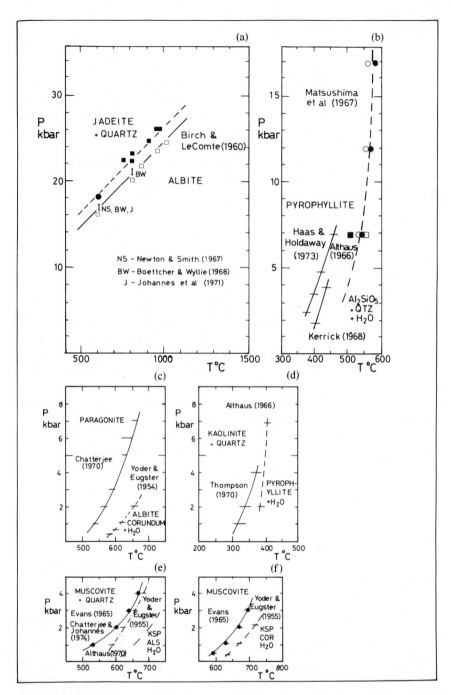

Fig. 1. Comparison of equilibrium, pressure–temperature curves for six independent equilibria that have been determined by methods involving reactants and reactants seeded with products. The results of the unseeded runs are shown with a dashed line, those of the seeded runs with a solid line. In each case it is apparent that a consistent amount of overstepping is required to give reaction where no seeds are initially present. (The magnitudes of overstepping observed are given in Table 1.)

the interfacial free energy (γ) and the free energy released on reaction (ΔG_r). The required overstepping will be smallest at high absolute temperatures, for reactions with large entropy or volume changes, and where the interfacial free energy of the new phase is small.

Brace and Walsh (1962) and Spry (1969, p. 148) give values of surface free energies for several minerals. These are obtained from measurements made in vacuum or air. Work in metallic systems (e.g., Porter and Easterling, 1981, p. 122) shows that interfacial energies in polygranular aggregates are about one-third the equivalent crystal surface energies. This suggests values of γ of 0.06–0.6 J m^{-2} for the major rock-forming minerals (see Appendix II).

Appendix II gives the overstep required for nucleation calculated from the extreme values given above, for, respectively, a high ΔS dehydration reaction and a low ΔS solid–solid reaction. The calculation assumes that nucleation takes place at grain corners (Cahn, 1956). It is suggested that nucleation takes place in general after 10–50 K or 1 kb overstepping of a reaction, the lower end of the temperature range for dehydration reactions, and perhaps up to 100 K where the reaction involves a small ΔS as in solid–solid reactions.

Experimental Evidence for the Amount of Overstepping of Mineral Reactions

Some measure of the amount of P or T overstepping is available from various experimental studies (Fig. 1, (a) to (f)). If nucleation rate is as critically dependent on the exact magnitude of overstepping as suggested by Eq. (1), then it will not be significantly affected by the heating or burial rate. The same number of nuclei will form between 10° and 11°C overstepping at a slow heating rate (e.g., 10°C m.y.$^{-1}$) as will form between 11° and 12°C overstepping at a heating rate 6 orders of magnitude faster (see also McLean, 1965, p. 105). It is this that allows the nucleation overstep observed in experiments to be considered applicable to real metamorphic rocks where heating rates are much slower.

Many early experiments used reactant assemblages unseeded with product phases (see Fyfe, 1960). There is a systematic ΔT or ΔP overstepping of an equilibrium boundary when experiments in which the initial assemblage was a mixture of both reactants and products are compared to those in which no seeds were used. In comparing experiments for solid–solid (Fig. 1(a)) to dehydration reactions (Fig. 1(b) to (f)) from various studies, we have necessarily assumed that the P–T reaction displacements are due to the nucleation barrier rather than due to starting materials of different composition, mode of preparation, or to experimental technique. An enormous P–T overstepping (about 6 kb at 1100°C) was observed by Boyd and England (1959) in their study of the reaction of aluminous enstatite + sapphirine + (sillimanite?) to pyrope. Uncertainties in the proportions and compositions of the reactants, as well as their thermodynamic properties prevents evaluation of the dependence of the amount of overstepping on ΔS_r, ΔV_r, ΔG_r. Other values obtained for Fig. 1 are shown in Table 1.

The difficulties of synthesizing Al_2SiO_5 polymorphs are well known, and most attempts to determine equilibrium boundaries in this system after the early studies (e.g., Roy and Osborn, 1954) were made with mixtures of polymorphs. The equilibrium boundary for kyanite–sillimanite is well established over a range of P–T conditions, but few data are available on the transformation of one polymorph to the other to enable an evaluation of the amount of overstepping. Birch and LeComte's (1960) location of the unseeded transformation of albite to jadeite + quartz compared to those of several workers using seeded mixtures (Fig. 1(a)) leads to values of an isobaric overstepping at $\Delta G_{1100 \ K}$ of 2300 J and an isothermal overstepping at $\Delta G_{16 \ kb}$ of 2000 J. Each of the dehydration reactions shown in Fig. 1(b) to (f) involves loss of one mole of H_2O, but the amount of overstepping varies from about 40° to 100°C. Some of the difference is clearly attributable to samples and techniques in both the unseeded and seeded experiments. Table 1 summarizes the results of the various comparisons possible. It can be seen that the ΔG_r overstepping required for nucleation covers a range of almost an order of magnitude. Some patterns are, however, discernible. The lowest overstepping is for reactions at higher temperatures or involving a reaction in which the reactants provide some epitaxial base for nucleation of the products (e.g., kaolinite $\rightarrow$ pyrophyllite). The values for ΔT overstepping given in Table 1 are likely to be maxima when compared with the behavior of real rocks. An experimental capsule provides a very restricted range of nucleation sites, whereas in natural systems the lowest energy site for nucleation may involve a phase not directly participating in the reaction (see Carmichael, 1969).

Thermodynamic Evolution of a Rock after Overstepping of a Reaction Boundary

Fig. 2 illustrates, by tracing schematically the evolution of chemical potential of one component, the progress of a reaction for which the equilibrium boundary was overstepped by a finite temperature before nucleation. The nature of this history is independent of the exact magnitude of overstepping. At equilibrium, the chemical potential of any component is the same in the product and reactant assemblages. After finite overstepping, but before any nucleation, the reactant assemblage is metastable with respect to the products. Nucleation under such conditions has the effect of producing a local volume of low chemical potential. From nucleation up until reaction completion there will be local chemical potential gradients in the rock. These promote the diffusion of ions necessary for a complex reaction involving several phases. They will be present whether the rate-controlling step in growth is diffusion or reaction at the grain interface (see below and Loomis, 1983).

Table 1. Estimation of the amount of overstepping required to give nucleation in some experimental studies

Reaction	Unseeded experiments	Full assemblage	Pressure (kb)	T Overstep (K)	G Overstep (J)
Pyp → Sil/And + 3Qtz + H$_2$O[a]	Althaus (1966)	Althaus (1966)	7	50	4500[b]
Pyp → Sil/And + 3Qtz + H$_2$O	Althaus (1966)	Haas and Holdaway (1973)	2	90	8000
Pyp → Sil/And + 3Qtz + H$_2$O	Althaus (1966)	Kerrick (1968)	2	70	6200
Kao + 2Qtz → Pyp + H$_2$O	Althaus (1966)	Thompson (1970)	2	45	1200
Par → Alb + Cor + H$_2$O	Yoder and Eugster (1954)	Chatterjee (1970)	2	90	4600
Mus → Ksp + Cor + H$_2$O	Yoder and Eugster (1955)	Evans (1965)	2	40	1450
Mus + Qtz → Ksp + Sil + H$_2$O	Yoder and Eugster (1955)	Evans (1965); Chatterjee and Johannes (1974)	2	95	3400
Mus + Qtz → Ksp + Sil + H$_2$O	Yoder and Eugster (1955)	Althaus et al. (1970)	2	70	2500

[a] Pyp-Al$_2$Si$_4$O$_{10}$(OH)$_2$: Sil/And-Al$_2$SiO$_5$: Kao-Al$_2$Si$_2$O$_5$(OH)$_4$: Qtz-SiO$_2$: Par-NaAl$_3$Si$_3$O$_{10}$(OH)$_2$: Mus-KAl$_3$Si$_3$O$_{10}$(OH)$_2$: Alb-NaAlSi$_3$O$_8$: Ksp-KAlSi$_3$O$_8$: Cor-Al$_2$O$_3$.

[b] Calculated at the temperature range of the experiments using entropy data for H$_2$O from Burnham et al. (1969) and for minerals from Robie et al. (1978): Par and Kao estimated.

AT EQUILIBRIUM AFTER NUCLEATION

DIFFUSION CONTROL

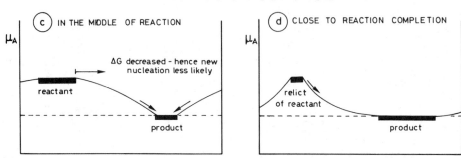

INTERFACE CONTROL

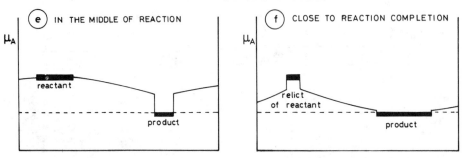

Fig. 2. Variation of the chemical potential of one component (μ_A) with position on a grain scale in a rock matrix at different times during the progress of a metamorphic reaction. The component could be Al_2SiO_5, and the reaction kyanite → sillimanite. It is assumed that before nucleation (a) the change in conditions is slow enough that no chemical potential gradients develop in the rock. This would be the case if there was an intergranular phase or fluid phase present within which diffusion was relatively rapid. Nucleation after a finite temperature overstepping (b) produces local volumes of low potential, hence driving diffusion. The rate of reaction after nucleation can either be diffusion controlled ((c) and (d)), or interface-reaction controlled ((e) and (f)). Note that after a finite amount of product growth, the chemical potential difference between product and reactant is reduced within a halo around the product. These "haloes" will be unfavorable sites for subsequent nucleation.

Effects of Heating and Burial Rates

It was shown above that the overstep required for nucleation is effectively independent of the heating rate in metamorphism. Faster heating or burial rates will, however, reduce the effect of reaction enthalpy (or entropy) and volume change on the local rock temperature or pressure (Ridley, 1985). If a reaction is initially overstepped, the effect of ΔH_r or ΔV_r is always to try to restore conditions back toward the appropriate equilibrium P and T during reaction progress. This effect is more likely to be important at slow heating or burial rates. The temperature or pressure may be restored early in reaction progress to a value at which nucleation no longer takes place and reaction rates are slower. Because of this, a fast heating or burial rate will in general give rise to more nuclei, and hence a finer grained product. In a complex rock in which there may be a small number of energetically favorable sites for nucleation (Yardley, 1977), a slow heating rate will make it more likely that nucleation occurs only on these sites. A faster heating rate favors more random nucleation.

Factors That May Influence the Overstepping Required for Nucleation

Undeformed pods of rock in an otherwise penetratively deformed metamorphic terrane often show clear evidence of disequilibrium, sometimes with the preservation of assemblages that have become thoroughly overprinted in the surrounding deformed rocks (Ridley and Dixon, 1984; Meyer, 1983). The absence of deformation appears to allow larger overstepping of reaction boundaries.

The catalytic effect of deformation on metamorphic crystallization has been examined by Dachille and Roy (1964) and Knipe (1981), among others, the most important effect probably being that deformation gives rise to high-energy defects in a mineral. These are favorable sites for the nucleation of a new phase (e.g., White, 1975).

The presence and composition of a fluid phase is also considered important, though it is difficult to separate the effects of these from those of deformation, since deformation and fluid influx into an originally dry rock often appear related (Beach, 1976; Heinrich, 1982; Rutter and Brodie, 1985). The absence of a fluid phase will substantially reduce the rates of grain-boundary diffusion, and hence of any grain-size scale diffusion in a rock (Thompson, 1983; Walther and Wood, 1984). In a rock in which chemical diffusion was restricted there may be significant chemical potential differences of any component between different points in the rock. The activity of Al_2O_3, for example, may be higher immediately adjacent to an aluminosilicate grain than along a quartz–quartz grain boundary. One important effect of this is that, where a reaction involves components not readily available in adjacent phases, a larger overstepping will be required to give a high enough chemical potential of all necessary components for a new phase to nucleate at one point in the rock (Fig. 3).

McLean (1965) suggests that the presence of chemical impurities in a rock may substantially affect the overstepping required for nucleation, because trace amounts of impurities are adsorbed along grain boundaries and reduce the grain interfacial energy. A fluid phase will act as an impurity in this sense (Fyfe *et al.*, 1958, p. 42). To what extent this can affect the overstepping required for nucleation is not clear.

The Kinetic Effects of Fluid Infiltration as Opposed to Temperature Change

It has been shown above that the nucleation rate law and the overstepping required for nucleation are the same whether a metamorphic reaction is caused by fluid infiltration or by temperature change. The two cases may, however, be different kinetically because of differing mechanisms of flow of fluid and heat. At present, the physical processes of fluid flow in metamorphic rocks (both infiltration and escape) are not well understood. Fluid influx could have the effect of giving rise to a very rapid, large overstepping of an equilibrium boundary—much more rapidly than could be achieved by heating or burial. The fluid released in a devolatilization reaction will influence the porosity of the rock. Rumble *et al.* (1982) describe a possible example of this in impure carbonates sandwiched between pure and massive carbonate horizons. Because of the uncertainty over the physics of such fluid flow, however, it is difficult at present to suggest how reaction textures produced when a reaction is prompted by fluid infiltration may be distinctive. Abnormally large grains, e.g., of garnet, in some skarns and marble borders to pelites, may be such examples.

Overstepping and the Possibility of Disequilibrium Crystallization

Overstepping of an equilibrium boundary before nucleation of a product mineral allows the possibility that this product will crystallize with a disequilibrium composition, or that a metastable mineral will crystallize in its place. Whether this phase is eventually preserved depends on subsequent reaction kinetics.

At present, there are no models for predicting the composition of a phase growing under disequilibrium conditions (cf. Loomis, 1983). Fig. 4 illustrates the range of possibilities for a divariant reaction, assuming that the only requirement is that the total free energy of the system is reduced. The range of possible disequilibrium compositions increases with increased overstepping. At a point approximately halfway across a divariant $(T–X)$ loop, the product may grow with the same composition as the original reactants. This

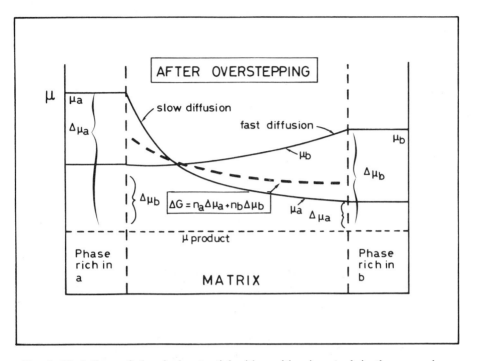

Fig. 3. Variations of chemical potential with position in a rock in the case where diffusion in the matrix is restricted. Imagine a reaction A + B → C, where phase A is rich in component *a* (showing very restricted diffusion), and phase B is rich in component *b* (showing less-restricted diffusion). The diagram shows schematically the variations in chemical potential expected for *a* and *b* after a finite overstepping of the reaction boundary. The free energy released on nucleation of phase C will be given by $\Delta G = n_a \Delta \mu_a + n_b \Delta \mu_b$, where $n_{a,b}$ is the stoichiometric coefficient of the component in the reaction, and $\Delta \mu_{a,b}$ the difference between the chemical potential of the component inside and outside the nucleus. It can be seen that ΔG is largest immediately adjacent to the phase richest in the slowest diffusion species (*a*)—i.e., coronitic growth is favored.

may be important if intracrystalline diffusion within the reactant phase is slow.

The crystallization of a metastable phase appears promoted where a product phase, sometimes one of two or more in a reaction, fails to nucleate. Nucleation kinetics favor the metastable formation of phases with higher entropy (Putnis and McConnell, 1980, p. 141). Matthews and Goldsmith (1984) systematically examined metastability in the reaction anorthite + $H_2O \rightarrow$ kyanite + zoisite + quartz. Kyanite failed to nucleate, and the actual reaction involved either mullite or margarite. A natural case of the failure of kyanite to nucleate (andalusite nucleating metastably in its place) has been described by Hollister (1969). A natural example where topotactic effects favored the crystallization of a metastable amphibole was described by Ridley and Dixon (1984).

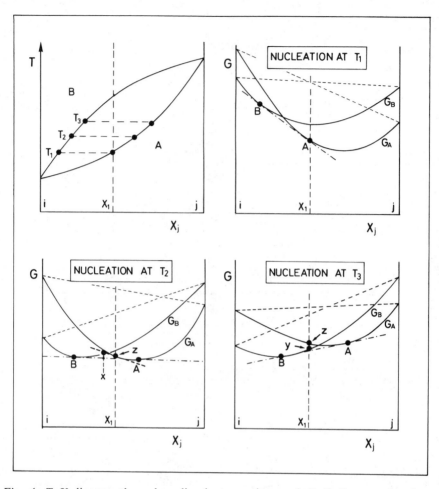

Fig. 4. T–X diagram through a divariant reaction, and G–X diagrams for three temperatures at which both solid–solution assemblages are stable. Consider a composition X. If nucleation of assemblage B occurs at T_1, the composition of B is constrained to be the equilibrium composition. If there is a finite overstepping of the equilibrium curve until T_2 before nucleation occurs, the total free energy of the system will be reduced if B nucleates with a composition x, or richer in component i'. A composition close to x is favored if intracrystalline diffusion within the reactant (A) is restricted. If nucleation takes place at T_3, then B can nucleate with the same composition as A: There will be a drop in total free energy (z–y).

Nucleation—Growth Models and Disequilibrium

Rate-Controlling Steps in Grain Growth

The rate of growth of a phase after nucleation is controlled by the slowest of three processes (Fisher, 1978): (1) diffusion to or from the growing or resorbing phase, (2) the rate of dissolution or precipitation at an interface, and (3) the rate of supply of heat needed to counteract the energy absorbed or released on reaction. The last possibility is not considered further here, because if overstepping before nucleation is significant, the energy required for reaction can be obtained by changing the rock temperature (Ridley, 1985).

The rate laws for diffusion-controlled growth and interface-controlled growth of spherical isolated nuclei are given in Appendix I. Both rates are predicted to increase with the magnitude of overstepping, interface-controlled growth approximately exponentially, and diffusion-controlled growth approximately linearly. Both laws also predict accelerating rates of reaction as reaction proceeds (Christian, 1975, p. 542).

The forms of Eqs. (2) and (3) of Appendix I assume rate control by processes at the product phase. In metamorphic reactions both reactants and products may be equally scattered through the rock, and the rate control may be given by dissolution of, or by diffusion away from, a reactant (Carmichael, 1979; Aagaard and Helgeson, 1982). If the rate is controlled by dissolution of the reactants, rates should decrease with reaction progress. This is expected to be more likely as reaction approaches completion, especially if one reactant becomes entirely enclosed within a new phase.

In any textural study, the rate-controlling step in a reaction can change with reaction progress (e.g., Fisher, 1978). Textures in a rock resulting from one specific rate control do not imply that reaction throughout was so controlled.

Absolute Rates of Reaction

There is little agreement in the literature as to what in general the rate-controlling step in a typical metamorphic reaction is (e.g., Loomis, 1976; Yardley, 1977; Fisher, 1978; Carmichael, 1979; Wood and Walther, 1983; Helgeson et al., 1984; Tracy and McLellan, 1985). Nor is there any agreement over likely absolute reaction rates.

Rates of growth of minerals in experimental studies, where these are likely to be independent of diffusion, have been analyzed by Fisher (1978) and Wood and Walther (1983) to give some indication of rates of growth in natural systems. These authors suggest that there is a universal rate law for growth or dissolution in silicate reactions, with rate dependent only on temperature. These rate laws predict growth of a typical metamorphic grain within a few hundreds of years.

A more thorough study of one specific system, that of feldspar dissolution, has been given by Helgeson *et al.* (1984). These authors show that, at least for this reaction, the reaction rate is generally strongly dependent on the pH of the fluid phase. They also emphasize that the rate of dissolution is dependent on the state of deformation of a grain, since this controls what proportion of the grain surface is "damaged" and hence more susceptible to dissolution. An analysis of the available experimental data, taking into account these factors, suggests significantly slower reaction rates at metamorphic temperatures than suggested by Wood and Walther (1983).

We suggest the following additional drawbacks in any use of such universal data to infer natural reaction rates. (1) The data compilation of necessity uses successful experiments, although in many similar experiments growth or dissolution has been reported as being very slow, especially where amphibole is involved (e.g., anthophyllite; Greenwood, 1963). (2) The experiments involve end-member phases, generally with a simple structure. The activation entropy for growth of such phases is likely to be large, and the growth rate fast, compared to a partially ordered phase of complex chemistry (Putnis and McConnell, 1980, p. 142; Carpenter and Putnis, 1985). (3) The experiments involve crystallization from a relatively large free volume of fluid, rather than at a tight grain boundary. The same kinetic processes may not be rate controlling in the two cases: Matthews and Goldsmith (1984) report reduced growth rates with smaller amounts of water in an experimental capsule.

Diffusion in metamorphic rocks is assumed to take place predominantly along grain boundaries. The rates of diffusion therefore lie between those in an open fluid (10^{-8}–10^{-9} m^2s^{-1}), and those within silicate phases (slower than 10^{-17} m^2s^{-1}). The few attempts at measuring sample bulk-rock diffusivities have been of limited success (e.g., Brady, 1979). Walther and Wood (1984) suggest that, at least during those periods of metamorphism when fluid is being released in reactions, bulk-rock diffusion on a millimeter to centimeter scale will take place along a fluid film coating the grain boundaries. Such diffusion will almost always be more efficient than true grain-boundary diffusion. Absolute rates will be related to the thickness of the fluid film. Walther and Wood's calculations suggest that a typical grain (radius > 1 mm) formed during prograde regional metamorphism would grow in 1000 to 10,000 yr. The normal case for their model would be of diffusion control rather than of interface-reaction control on absolute rates. Significantly slower rates of growth have been inferred for a rock that is not actively undergoing dehydration (see Carlson and Rosenfeld, 1981).

Causes and Effects of Varying Growth Rates and the Relations between Nucleation and Growth Rates

Some aspects of the effects of differing growth rates in metamorphism have been discussed by Spry (1969, pp. 122–136). Faster growth rates lead to a greater likelihood of a coarse grain size and also to the growth of poikiloblastic grains. Obviously, a greater overstepping of a reaction will lead to faster growth rates, all other factors being equal. Most of the factors, however, that allow a greater reaction overstepping through the suppression of nucleation will also significantly affect growth rates. If diffusion takes place predominantly along grain boundaries, then the absence of a fluid phase will significantly reduce rates. Deformation increases the rate of mass transfer both by physically transposing volumes of different composition within a rock and also by producing pressure gradients on a grain scale (by the opening and closing of submicroscopic cracks) that will drive fluid movement along the grain-boundary network (Etheridge *et al.*, 1983). Reactions in a dry, undeformed rock will not only show greater overstepping but will also take place slowly because of restricted diffusion.

Larger overstepping of a reaction before nucleation favors coarse grain sizes and also a diffusion-rate control of growth over interface reaction control (Fig. 5). The diffusion rate increases less steeply than the interface-reaction rate with increasing overstepping. A change from one rate control to the other has been suggested from the results of experiments involving crystallization from a melt with varying amounts of supercooling (Kirkpatrick *et al.*, 1979), the change taking place in the system studied at 40–50°C undercooling. Certain related patterns can be predicted for metamorphic rocks. Because epitaxial nucleation will reduce the amount of overstepping required, it is more likely to result in an interface-reaction controlled texture than is unassisted nucleation. There may be, in some instances, an evolution from diffusion control to interface control during the growth of a phase in a divariant reaction—later growth taking place at small degrees of overstepping on already nucleated grains (Figs. 4 and 8).

Walther and Wood (1984) discuss some aspects of the importance of the amount of fluid in a rock in controlling reaction rates. They stress the effects of the mode of fluid release during a dehydration reaction in controlling this amount—how much fluid release takes place along discrete channels. If the thickness of the fluid film along a grain boundary is constant, then the bulk-rock diffusion is inversely proportional to the average grain size of a rock. A finer grained rock might show evidence of more rapid grain growth. It is useful, therefore, to consider the causes of overall grain (matrix) coarsening during metamorphism. The driving force for grain coarsening through the reduction of net interfacial free energy is small. It is considered to be potentially significant in monomineralic aggregates but not in a polyphase aggregate where coarsening requires dissolution and diffusional steps as well as grain-boundary migration. Even in monomineralic aggregates, coarsening may be inhibited by small amounts of impurities, e.g., graphite along grain boundaries (Robinson, 1971). Grain coarsening in prograde metamorphism appears more generally to be promoted when the phases forming the rock matrix become involved in bulk-rock metamorphic reactions, either directly or indirectly if a reaction cycle (Carmichael, 1969) involves the resorption of a phase at one point and its reprecipitation elsewhere.

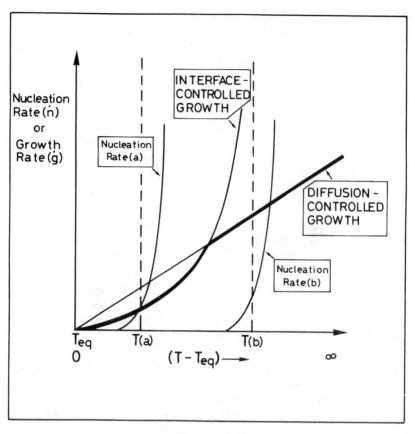

Fig. 5. Diagram to illustrate the relations between the rates of diffusion, interface reaction, and nucleation with increasing temperature of overstepping. If nucleation kinetics are such that nucleation reaches a significant level at $T(a)$, growth will be interface controlled. If nucleation requires a larger overstepping (up to T(b)), growth will be diffusion controlled. This pattern will be the same whatever the relative ease of diffusion as compared to surface reaction. The thickened line shows which mechanism is dominant for a particular value of $(T - T_{eq})$.

Grain-Size Distributions, Nucleation-Growth Models, and Disequilibrium

An analysis of the distribution of grain sizes of a certain mineral in a rock, especially if the grain sizes are regarded as resulting solely from a single crystallization event, should give some information about the processes of nucleation and growth. The most comprehensive studies undertaken have been by Kretz (1973, 1974), using both the grain size and composition to trace nucleation and growth. The spatial distribution of grains shows that the growth of each grain was effectively independent of those surrounding.

This is a very blunt method of tracing nucleation and growth. The method will work quantitatively so long as (1) the rate of growth is dependent only on the size of a grain (and absolute time); (2) the composition of the growing grain is always an equilibrium composition or everywhere equally removed from equilibrium; and (3) there is no intracrystalline diffusion or resorption after growth. How much the conclusions are affected by such possible complexities as variable growth rates depending on environment, time-dependent growth rates, and diffusion, is less clear.

The qualitative conclusions of the work of Kretz and similar earlier studies (Jones and Galwey, 1964, 1966) are that (1) with time, nucleation rate generally increases and then decreases, though may sometimes be multimodal, and (2) reaction rates increase quite strongly with time until reaction is almost complete. The study of Kretz (1973) suggests that growth is proportional to the square root of the grain radius, a law that is most consistent with diffusion-controlled growth (Christian, 1975, p. 542).

Perhaps the most important conclusion from these studies is that nucleation is not instantaneous, but continues in an apparently random fashion almost to reaction completion. In the case studied by Kretz there is an increase in nucleation rate up until reaction almost reaches completion. Such histories are consistent with progressive heating through a reaction isograd, with a cutting out of nucleation either because of a slight reduction in temperature as energy is absorbed in reaction (Ridley, 1985) or because of a reduction in reaction affinity as the whole matrix becomes depleted in reactants (B. Yardley, 1984, personal communication).

A contrasting aspect of modeling grain growth with a consideration of kinetics is discussed by Loomis (1982) and Loomis and Nimick (1982). These studies consider a certain compositional system and trace the zonation of garnet, the longest surviving phase in the system considered, through several reactions. The general predictions are that zonation, after initial overstepping and with interface-controlled growth, should be similar to that of equilibrium crystallization, but of slightly lower magnitude. If overstepping invariably occurs, and hence there is no possibility of any comparison between natural examples of disequilibrium and equilibrium zonation profiles, the use of zonation pattern to give information about kinetics in the way envisaged by Loomis relies on the reliability of the thermodynamic data used in modeling growth.

Specific Reactions, Specific Textures, and Their Relationships to Disequilibrium Processes

The actual impetus for textural development, the growth of a certain mineral at a certain place, is provided by the free energy difference between reactants and products after the overstepping of a reaction boundary. In the following sections, examples of textures, both specific natural examples and general classes, formed in various types of reactions (devolatilization, poly-morphic, solid–solid heterogeneous) are considered in the light of the kinetic principles outlined above.

Devolatilization Reactions and the Development of Porphyroblasts

Dehydration reactions in metapelites and amphibolites and decarbonation reactions in metacarbonates are probably responsible for most of the major mineralogical changes that take place during the metamorphic history of a typical terrane.

There are certain general features of devolatilization reactions that will affect their kinetic behavior. Such reactions have a much larger reaction enthalpy (entropy) than solid–solid reactions, hence it is anticipated that the ΔT overstepping required to allow nucleation will be relatively small (see above). Because such reactions have large reaction enthalpies, a small amount of reaction may have a significant effect on the rock temperature: reducing it soon after the onset of reaction to a level at which no further nucleation takes place (Ridley, 1985, and in preparation). The possible ef-fects of the fluid released by reaction on subsequent kinetic behavior have been discussed by Walther and Wood (1984). This fluid would be expected to "saturate" grain boundaries and hence allow relatively fast diffusion be-tween reactants and products.

Devolatilization reactions appear to favor porphyroblast growth. Because most field description and interpretation of isograds are based on the occur-rence of porphyroblasts as "index minerals," it is useful to consider specifi-cally kinetic and chemical aspects of their formation and behavior. Growth of a porphyroblast is apparently related to the kinetics of metamorphic crys-tallization (Atherton and Edmunds, 1966; Spry, 1969, p. 138). The growth of a phase as a porphyroblast requires a low nucleation rate to growth rate ratio $(\dot{n}/\dot{g})$, or more strictly, low cumulative nucleation through the process of a reaction.

A low nucleation rate is favored by a high interfacial energy of the new-grown mineral, or by slow rates of change of temperature, pressure, or fluid composition. Nucleation may take place over only a short interval of reac-tion progress if diffusion is rapid and depletion haloes develop rapidly around growing grains, or if the reaction has a high ΔS and the temperature becomes rapidly buffered. Rapid growth will be favored by high diffusion rates. These can be the result of a fine matrix grain size or a large amount of intergranular fluid in the rock (Walther and Wood, 1984).

Eskola (1939) noted that, despite obvious exceptions, Becke's (1913) idio-blastic series provided a good indication of which minerals will tend to form porphyroblasts and which will tend to form the matrix. Kretz (1966) suggested that the position of a mineral in this series is related to the specific interfacial energy of a mineral crystal form. This is simply an observation of the first criterion for low nucleation rates stated in the previous paragraph—the effect of high surface energy. It can be seen from Eq. (1) that, if the interfacial energy of the product is higher, a larger overstep of the reaction equilibrium is required before nucleation. (In a devolatilization reaction, this will counteract the effect of a high reaction enthalpy on the magnitude of this overstep.) For two reasons, a large overstepping before nucleation increases the likelihood of the reaction enthalpy controlling temperature early in reaction progress. A larger overstepping means that growth rates, whether controlled by diffusion or interface processes, will be faster. The combination of a large overstepping and large ΔS_r means that the nucleation rate increases less steeply with increasing temperature, and a longer time is therefore required at a given rate of heat input to reach a temperature at which nucleation not only occurs, but is also fast.

Evolution and Sequential Growth of Porphyroblasts

Loomis (1982, 1983) examined a hypothetical prograde reaction sequence for a metapelitic rock in which garnet first appears, becomes partially resorbed, and then later undergoes further growth (see also Thompson et al., 1977). Loomis suggests, however, that the resorption of garnet would not in reality occur, because the garnet would act as a refractory phase and once formed not partake in any reactions.

The prograde reactions forming the sequential porphyroblast phases in a typical pelite sequence should in many cases involve the resorption or disappearance of earlier formed porphyroblasts (e.g., Thompson et al., 1977). In many instances, this is not apparent from textural evidence. The evidence is, rather, that each sequential porphyroblast grew directly from the matrix phases (usually chlorite + biotite + muscovite + quartz + albite; Chinner, 1967; Atherton, 1965, 1976; Fox, 1975). Staurolite, for instance, appears often to grow through the AFM continuous reaction:

chlorite + muscovite → staurolite + biotite + quartz + H_2O ± $FeMg_{-1}$,

rather than the AFM discontinuous reaction (Thompson, 1976):

garnet + chlorite + muscovite → staurolite + biotite + H_2O ± quartz,

despite the fact that the second reaction should be reached at lower temperatures (Fig. 6). For an initial biotite–chlorite–muscovite–quartz assemblage with a composition X (Fig. 6), garnet should first appear at T_1. Because of the overstepping required for nucleation, it may not appear until T_2, though this will have little effect on the rock evolution. Staurolite should first appear at the univariant reaction. If, however, nucleation is delayed until T_3, chlo-

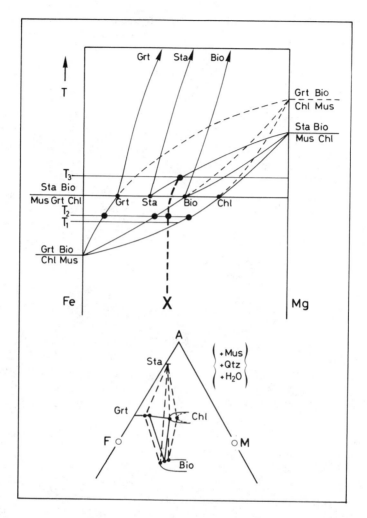

Fig. 6. *T–X* loop and AFM projection for the phases Grt-Sta-Bio-Chl-Mus (after Thompson, 1976), simplified to show only those reactions relevant to pelitic compositions less Al rich than the chlorite–garnet join. The AFM projection shows the situation at reaction at T_3. (Solid lines are metastable assemblage, dashed lines stable assemblages.)

rite + muscovite could break down to form staurolite without significantly affecting the modal garnet in the rock.

The case for the alumino-silicates is perhaps clearer. Pitcher (1965, p. 333) and Spry (1969, p. 272), among others, have commented that there is rarely evidence to suggest that coexisting andalusite–sillimanite–kyanite are actually in equilibrium. The relatively frequent occurrence of at least two of these polymorphs together suggests that even where there is no nucleation barrier the relevant reactions can still be very slow. The sequential growth of sillimanite following kyanite appears to be due to reactions such as (Charmichael, 1969):

muscovite + staurolite + quartz → biotite + sillimanite + H_2O,

or (Powell, 1978, p. 166–168):

muscovite + quartz → K-feldspar + sillimanite + H_2O,

rather than the polymorphic reaction:

kyanite → sillimanite.

Carmichael (1969) provided an explanation that seems to apply to many rocks of why porphyroblasts rarely directly replace each other. Quirks of nucleation and differing diffusion rates for different ionic species mean that it will often be "easier" for a reaction between porphyroblast phases to take place through the dissolution and reprecipitation of the matrix phases, which effectively act as catalysts. The resorbant porphyroblast will be rimmed or replaced by phases not necessarily partaking significantly in the reaction taking place in the rock as a whole (e.g., plagioclase in the case discussed by Carmichael, 1969, for the reaction staurolite + muscovite + quartz → garnet + biotite + sillimanite + H_2O). This is not, however, regarded as an explanation for every instance where a porphyroblast once formed appears reluctant to react. The rate of growth and the rate of dissolution of any grain should be approximately the same for a similar reaction affinity (Wood and Walther, 1983), assuming that the rate-controlling step in the reaction and the mechanisms thereof are the same. A change in mechanism is possible in continuous reaction with phases of variable composition. A garnet-consuming reaction, for instance, will involve simultaneously a reduction in the modal proportion of garnet and a change in the composition of that remaining. Intracrystalline diffusion within garnet is slow (Freer, 1981). It is possible, therefore, that the progress of any continuous reaction involving the breakdown of garnet will be rapidly slowed once a narrow rim of the resorbing grain reaches equilibrium with the matrix (e.g., Loomis, 1975). No such reaction control is possible during garnet growth. Similar considerations presumably apply to other porphyroblast phases that show zoning. The importance of the resistance of porphyroblasts to deformation is clear in this respect. Stress-induced recrystallization by the movement of grain or subgrain boundaries through a phase will allow much more rapid equilibration of the phase than intracrystalline diffusion (Ridley and Dixon, 1984).

Aspects of the kinetic control on the change of porphyroblast composition are discussed further below with regard to their application in geothermometry and geobarometry.

Relative Grain Sizes of Sequential Porphyroblasts

Fig. 7 shows two examples of garnet–staurolite schists with similar mineralogies (quartz + muscovite + biotite + garnet + staurolite + albite + ox-

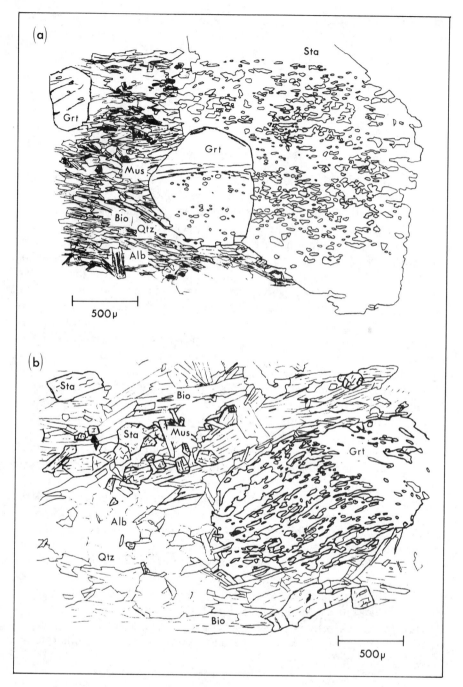

Fig. 7. Two examples of Grt-Sta-schists showing markedly different porphyroblast textures. (a) is from northwest Maine (the upper-staurolite zone of Guidotti, 1970b). (b) is from the staurolite zone of Connemara (Yardley *et al.*, 1980). Scale bar is 500 microns long.

ides), but with contrasting textures. That from Maine shows euhedral garnet and staurolite, the staurolite significantly coarser grained than garnet. The sample from Connemara shows xenoblastic, ragged garnets and fine-grained staurolite of grain size not significantly coarser than the matrix biotite and plagioclase.

There is a marked difference in matrix grain size between the two rocks—that from Connemara being on average almost ten times coarser. The sample from Connemara also shows evidence of a period of significant matrix coarsening between garnet and staurolite growth (cf. Yardley *et al.*, 1980): The garnet inclusion trail shows that it overgrew a quartz–phyllosilicate significantly finer grained than that seen at present. The differences in porphyroblast texture are therefore consistent with, though obviously not proof of, a model in which the size of a porphyroblast phase is related to the ease of diffusional mass transfer in the matrix.

Polymorphic Solid–Solid Reactions

The growth of one polymorph from another, even where this involves a complete reconstruction of the lattice, will involve different kinetic controls than if a complex dehydration reaction produces the same phase. If a polymorphic reaction takes place by direct replacement, no diffusion or transport step is involved, and the reaction rate is of necessity interface controlled. It is not, however, invariably the case that polymorphs directly replace one another. The other kinetically important factor of polymorphic reactions is that they generally involve small entropy changes. Large temperature overstepping of an equilibrium boundary before nucleation is therefore more likely, and absolute reaction rates may be slow because of small driving energies. Two sets of polymorphic transitions have received much attention in field and laboratory studies: $CaCO_3$ and Al_2SiO_5.

The $CaCO_3$ calcite to aragonite transformation has been studied experimentally to determine what the preservation of aragonite in blueschists terranes implies about P–T conditions or rates of exhumation. Unfortunately, an examination of the various experimental investigations of this transformation (Rubie and Thompson, 1985) does not reveal any simple kinetic criteria that may be easily applied to the understanding of the reaction in nature.

Al_2SiO_5 Polymorphs

The widespread distribution of the Al_2SiO_5 polymorphs in contact and regional metamorphic environments has resulted in many attempts to deduce P–T conditions of metamorphic equilibration and P–T–t evolutions from their occurrence.

In regional metamorphism, the replacement of one Al_2SiO_5 polymorph by another appears rare. The most common direct replacement appears to be fibrolite after andalusite, as deduced by chiastolite-like shapes and graphite inclusion trails in fibrolite aggregates (Fowler-Billings, 1949, p. 1262; Atherton, 1965; Rosenfeld, 1969; Tracy and Robinson, 1980), though kyanite after andalusite has also been reported (Chinner and Heseltine, 1979). In contact metamorphism, replacement textures among the polymorphs are more common, e.g., kyanite by sillimanite (Naggar and Atherton, 1970), andalusite by sillimanite (e.g., Loomis, 1972), and also andalusite by kyanite (Hollister, 1969). These patterns and the relatively frequent coexistence of two alumino-silicate polymorphs in a rock (much more frequent than predicted from equilibrium considerations; see Strens, 1968) suggest that kinetic factors are important in determining reaction textures among these polymorphs. Hollister (1969) further suggests that the polymorphs may nucleate and grow outside their stability field. It is necessary to consider, therefore, what influences the nucleation site and subsequent growth in any case.

It should be possible to make some inferences about actual kinetics from an examination of patterns of differences in perceived reaction progress between cases where polymorph replacement is direct, and where it is indirect. Hollister (1969) notes that in an inferred sequence of direct replacements, andalusite → kyanite → sillimanite, the andalusite to kyanite replacement is always seen complete, whereas kyanite is frequently seen as relicts within sillimanite. It remains to be demonstrated whether (1) the time interval for the andalusite to kyanite transformation was longer than that for kyanite to sillimanite, (2) any similarity of the andalusite and kyanite structures allowed a reaction pathway faster than complete dissolution and new growth of the replacement polymorph, or (3) differing reaction rates were related to differences in the nature of the interfaces (e.g., impurity content) in the two transformations.

In the reaction of kyanite to sillimanite, it appears that incomplete replacement is more common when the reaction is indirect (e.g., through the mechanism described by Carmichael, 1969), rather than direct. This suggests that reaction rates are slower in the indirect replacement, and hence that there is a diffusional control on reaction progress. Fisher (1970), however, considered that the overall reaction rate is always controlled by the dissolution or growth of one of the alumino-silicate phases. This possibility is not ruled out by the observations of the frequency of partially completed reactions. Nucleation of the new alumino-silicate phase often takes place on very specific sites (sillimanite grows parallel to the [010] directions of biotite; Chinner, 1961; Yardley, 1977). These are presumably the first sites to be activated during progressive overstepping of the reaction boundary. The fact that the product has not also nucleated directly on the reactant, even though reaction did not go to completion, suggests that overstepping never reached a level that would allow such nucleation. Where replacement was indirect, reaction took place at lower degrees of overstepping, hence both diffusion rates and interface-reaction rates would be relatively slow.

Heterogeneous Solid–Solid Reactions

The high-pressure breakdown of plagioclase involves the heterogeneous decomposition of albite to jadeite + quartz, and of anorthite to grossular + kyanite + quartz. The reaction kinetics of such a decomposition will therefore be different from that of a polymorphic inversion, because more than one phase is required to nucleate and some diffusion is necessary during transformation. The kinetics of the reverse reaction will also be different as one phase forms at the expense of two or three. These reactions are treated separately from heterogeneous dehydration reactions, in part because the resultant reaction textures are often quite distinctive, and in part because of their importance in geobarometry and geothermometry.

Compagnoni and Maffeo (1973) describe, from eclogite-facies rocks, fine-grained pseudomorphs of intergrown jadeite + quartz + zoisite after plagioclase. This is a "eutectoidal" decomposition. Koons *et al.* (in preparation) point out that the pseudomorphing assemblage is not part of the equilibrium assemblage for the rock as a whole. The rock is effectively undeformed. Reactions between plagioclase and the other components of the rock (biotite, phengite, garnet) have been overstepped, presumably because of restricted intracrystalline diffusion. Reaction did not take place until the absolute stability limit of plagioclase itself had been overstepped. Jadeite + quartz aggregates after albite are seen in low-grade blueschist-facies rocks (e.g., Bloxam, 1960) but are rare in higher-grade rocks that show evidence of penetrative deformation. In these, albite presumably broke down earlier through a series of continuous reactions.

A typical texture of the reaction in the reverse direction is the growth of plagioclase along grain boundaries between garnet, kyanite, and quartz. This is described in more detail in the following section.

Solid–Solid Reactions and Geobarometry

Two solid–solid equilibria involving minerals allowing substantial crystalline solutions that are often used as geobarometers are:

$$\text{garnet} + \text{sillimanite} + \text{quartz} \rightarrow \text{cordierite},$$

and

$$\text{garnet} + \text{sillimanite/kyanite} + \text{quartz} \rightarrow \text{plagioclase}.$$

These are considered in detail here to show possible relationships between kinetics and geobarometric and geothermometric determinations, with the recognition that the two reactions often take place simultaneously.

The two reactions have similar characteristics. Because of crystalline solutions, the full four-phase assemblage is stable over a wide range of P and T conditions for most relevant rock compositions. The mineral on the right-hand side of the reaction as written is stable at lower pressures, hence the four-phase field is in many instances reached late in the metamorphic history of an area, along the decompression segment of a P–T–t path (Thompson and England, 1984; Thompson and Ridley, in preparation). The reactions, in both cases, form one phase at the expense of three and will always require concurrent ion exchange within the remaining garnet.

Where the reaction has gone to the right, a certain texture is characteristic of both assemblages. For the second reaction, the plagioclase may form narrow, "coronitic" rims between the alumino-silicate and quartz, completely enclosing the former. Cordierite, likewise, often forms a reaction rim between sillimanite and quartz, or garnet and sillimanite (Hollister, 1977; Loomis, 1979). Coronitic reaction textures indicate that nucleation of the new phase could only take place where at least one of the components required was directly available from a reactant mineral. This suggests that before nucleation there were gradients in the activities of species in the intergranular phase. Such gradients could develop with changing P–T conditions if intergranular diffusion rates only allowed continuous equilibration of the fluid phase with minerals that were in direct contact.

There is disagreement over whether rocks showing such textures and assemblages can be reliably used as geobarometers, and in the case of garnet + cordierite additionally as a geothermometer (Richardson, 1974; Hollister, 1977; Tracy and Richardson, 1978). The problem is essentially one of determining what parts of the assemblage were in equilibrium at any one point in the history.

In an equilibrium history for garnet + cordierite, the latter should nucleate at a pressure and temperature dependent on the Ca-Fe-Mg content of the garnet. With subsequent decreasing pressure, cordierite should grow at the expense of garnet + sillimanite + quartz, with the remaining garnet becoming richer in Ca and Fe. If zoning is preserved, cordierite should show growth zoning and garnet a form of reaction zoning (Loomis, 1975). The history of the garnet–plagioclase system would be similar, with the extra complication of parallel reactions providing the albite component of the growing plagioclase. The reaction zoning of garnet in this case would be toward Ca-poorer rims.

The potential problems of using these assemblages as geobarometers in rocks in which equilibrium between all the phases was not maintained can be illustrated with G–X diagrams (Figs. 4 and 8). The major possible barriers to an equilibrium history are the nucleation step, diffusion within the resorbing garnet, and diffusion through any coronitic rim composed of the reaction products. After overstepping, from mass-balance considerations, it can be seen that slow diffusion rates within garnet will favor the growth of cordierite with an Fe/Mg ratio shifted from equilibrium toward that of the garnet (Fig. 8). Such a shift will continue throughout growth. This history is different from that predicted by Loomis (1982) for the chlorite → garnet reaction, in which the only barrier to equilibrium was the nucleation step.

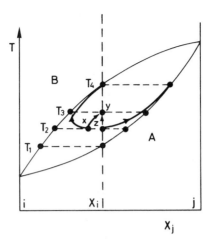

Fig. 8. Possible evolution of product and reactant compositions in a divariant reaction for which an overstepping was required to nucleate the product assemblage. Phase B nucleates at T_2 with composition x. If the dominant barrier to equilibration after nucleation is diffusion in the reactant, then the compositions of the two phases will evolve along the lines $x \rightarrow y$ and $z \rightarrow y$ with reaction completion at T_3. This would represent an extreme case. In the general case the bulk composition of the two phases will evolve approximately as indicated by the thick lines between T_2 and T_4: The exact evolution is dependent on the dissolution and growth rate, heat supply, and diffusion steps.

The effectiveness of a complete coronitic reaction rim in slowing down or sealing a reaction has been demonstrated in experimental studies of reaction rates by Kridelbaugh (1973) and Brady (1979). After the formation of a rim, subsequent reaction progress is determined by the rate of diffusion of one component through the barrier. The rate will decrease with increasing thickness of the rim. It is possible, therefore, that the bulk of the reaction takes place in a short time interval between the onset of nucleation and the sealing of the coronitic rim.

After sealing of the bulk-rock reaction, ion exchange may continue independently at the two interfaces of the corona (see Lasaga *et al.*, 1977). There is evidence for such Fe and Mg exchange in the garnet + sillimanite + quartz $\rightarrow$ cordierite reaction in the data given by Hollister (1977) and Loomis (1979). Both these studies report that a cordierite which has grown a significant distance away from any garnet grain has a lower Mg/Fe ratio than one immediately adjacent to garnet. This is consistent with a model in which continued equilibration with changing temperature took place longer at grain–grain contacts than between reacting grains separated by the matrix.

A combined geothermometric/geobarometric determination using the garnet + cordierite coexisting with sillimanite + quartz is normally made using compositions of immediately adjacent garnet and cordierite. If the phases maintained only local equilibrium, the temperature and pressure determinations may be recording the closure of two independent systems—the pressure determination reflecting an earlier stage in the rock evolution. The determination of $P–T–t$ paths using measured phase zonation in this reaction may not be more accurate than simple inferences drawn from the presence of the reaction textures.

Kinetic Information Deduced from Mapped Metamorphic Sequences

Zones of Persistence or "Smeared-Out" Isograds

Zones in a metamorphic terrane to the high-grade side of an isograd marking a nominally univariant reaction in which both the low-grade and high-grade assemblages are observed ("zones of persistence") have been described by various authors.

These zones have been explained assuming equilibrium conditions:

1. No reaction is perfectly univariant; trace components will always be distributed so as to give a divariant field, however narrow (e.g., Ramberg, 1952).
2. Any reaction that involves a fluid phase may become divariant. If the rocks effectively act as a closed system, then the activities of the fluid species are buffered by the reaction itself (Thompson, 1955; Guidotti, 1970a).

Both of these explanations could in many instances be tested. It is clear, however, that "zones of persistence" could also be the result of a kinetic control on the overall rate of reaction (Carmichael, 1979). Carmichael, from studying the variation in width of such zones, the effects of trace components, and the degree of "textural equilibrium," suggested that their characteristics are consistent with the overall rate of the reaction being controlled by the dissolution rate of one of the reactants.

The potential, but also the potential problems, of using the characteristics of such zones to give information about reaction kinetics can be illustrated by considering possible likely $P–T–t$ paths for rocks close to an isograd (Fig. 9)—a classical isograd based on the absence or presence of the higher grade assemblage.

There will be two reaction curves in $P–T$ space—the equilibrium curve, and, at higher temperature, the conditions required for nucleation. There will be a layer of rocks that reached the temperature of stability of the higher grade assemblage, but not a high enough temperature to give any nucleation. These rocks will not show any evidence of reaction. The rocks at the mapped isograd will be those that just reached the temperature required for nucleation. Nucleation will have taken place over a restricted length of time. There will, however, be a significant period before these rocks cool back beneath the reaction isograd (path x–y, Fig. 9). All nuclei should therefore grow to a finite size, i.e., the reaction overstepping allows a finite amount of reaction to take place effectively right at the mapped isograd. A certain distance above the isograd enough nuclei will have formed and the rocks will have spent sufficient time within the $P–T$ field of the product assemblage for the reaction to go to completion. If there are sufficient constraints on the $P–T–t$ paths of a suite of rocks (obtained from geobarometry, geothermometry, and radiometric dating), it would be possible to obtain information about reaction kinetics from studies of (1) the variation in reaction progress with distance from a mapped isograd; (2) variations in where a grain nucleates; and (3) grain size and grain-size distributions across "zones of persistence." There are several possible complications, e.g., changes in rock temperature because of energy absorbed during reactions, or changing fluid activities from absorption of fluid in continuous reaction during any interval of cooling (Yardley, 1981).

Information from Specific Microtextures

In principle, textures can give important information about metamorphic histories. In practice, however, we have too little control of what possible correlations there are between textures and processes to be able to identify more than qualitative aspects.

One recurring problem is the understanding of what determines the nucleation site of a new phase in any one instance. Some specific epitaxial relationships have been established (e.g., sillimanite on biotite, Chinner, 1961; or sphene on ilmenite, Spry, 1969, p. 200), but no concept of general patterns of preferential nucleation sites for each type of silicate structure exists. Garnet at the garnet isograd appears often to grow preferentially within quartz aggregates, rather than directly replace chlorite (Atherton, 1964; Carmichael, 1969). The exact reaction pathway followed in a system of linked subreactions as described by Carmichael (1969) is a function of the kinetics of nucleation and all the steps involved in dissolution and growth. In the examples described by Carmichael, it appears that the immobility of Al effectively controlled the textures produced. In other examples (e.g., Kwak, 1974; Foster, 1981, 1983), this particular constraint does not appear to have been important. Foster infers that the exact nucleation site of a product can determine both eventual textures and the assemblages pseudomorphing a reactant phase.

These are examples in which specific inferences about kinetic processes have been made from rock texture. Microtextures are used subconsciously in the determination of the direction of a reaction in a rock showing a

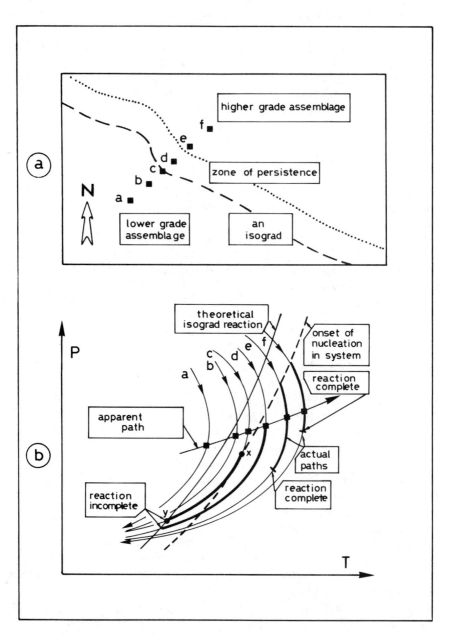

Fig. 9. Possible relations between pressure–temperature–time paths for six samples across an isograd, and reaction kinetics and the presence of a zone of persistence within where both the high-grade and low-grade assemblages coexist. (a) Schematic field map, showing the localities of the six samples relative to the mapped isograd and zone of persistence. (b) *P–T–t* paths for rocks from each of the six localities. The thickened lines indicate the time periods over which the prograde reaction took place, and show, schematically, after what overstepping reaction would be complete. The apparent path is the *P–T* geotherm that would be inferred if metamorphism was interpreted as being the result of proximity to a heat source at depth.

reaction assemblage, hence allowing an inference about the sense of change of pressure or temperature.

Certain patterns of microtextures can, however, indicate other aspects of the rock history or may be useful as an aid to interpreting chemical data; for example, over what scale a rock maintained chemical equilibrium. Fine-grained, isochemical, "eutectoidal" pseudomorphs formed of two or more phases after one (e.g., zoisite + quartz + jadeite after plagioclase as described above) appear to be restricted to rocks in which deformation was not penetrative and in which diffusion was restricted. Possible reactions involving this phase and the others in the rock have been overstepped until the breakdown reaction of the single phase occurs. This is the situation most likely to give rise to the crystallization of disequilibrium phases or phase compositions. A similar example, where a porphyroblast phase reacted with its inclusions and not with the matrix phases has been described by Vernon (1978); in this case in a retrograde metamorphism.

Such textural information may provide additional information on metamorphic (P, T, X_{fluid}) histories. At present, however, our knowledge of the exact processes involved in metamorphic crystallization is not complete enough to provide such information except in individual cases. More experimental data are needed on reaction kinetics to allow a determination of what the rate-controlling step for different reactions in different environments might be. More information is also required on where and when specific textures are found in metamorphic terranes. Are there any patterns of variation in textures of a specific phase in a prograde sequence? For example, Yardley et al. (1980) described differences in porphyroblast textures from a single lithological unit where it is repeated at different levels in a structural pile. Are such differences the result of local effects (e.g., rock and fluid chemistry, or deformation), or of regional effects (P–T–t paths or the rate of heating)? Are there consistent differences in textures of a specific phase between terranes in which metamorphism took place in different tectonic environments? Are there any correlations between textures and inferred high fluid-to-rock ratios?

Acknowledgments

The "work" for this contribution was undertaken while J.R. was in receipt of a Royal Society European Programme Research Fellowship. We wish to thank K. Malmström for typing and drafting; C.T. Foster, D.C. Rubie, and B.W.D. Yardley for scientific discussion; and D.M. Carmichael, J.M. Ferry, and C.T. Foster for helpful reviews of the manuscript. B.W.D. Yardley is also thanked for the loan of a number of samples, one of which is illustrated here. The sample illustrated from Maine was collected with C.V. Guidotti.

Appendix I

Fundamental Equations of Nucleation and Growth Rates

1. The nucleation rate per mole for a spherical nucleus in a homogeneous medium (after McLean, 1965, p. 105) is given by:

$$\dot{n} = A_n \exp\left(\frac{-16\pi\gamma^3}{3\Delta G_r^2 \kappa T}\right), \tag{A1}$$

where A_n is a constant given by:

$$A_n = \exp(\Delta G^*/\kappa T)N\nu \tag{A2}$$

(see Christian, 1975, p. 441), where ΔG^* is free energy of activation, N is Avogadro's number, ν is atomic vibration frequency, γ is interfacial free energy, T is absolute temperature, κ is Boltzmann's constant, and ΔG_r is the free energy released on reaction per mole.

An approximate relationship between the nucleation rate and the over-stepping of an equilibrium boundary can be obtained by noting that for small departures from equilibrium:

$$\Delta G_r = -\Delta S_r\,(T - T_{eq}), \tag{A3}$$

where T_{eq} is the equilibrium temperature, and hence:

$$\dot{n} \propto \exp[(T - T_{eq})^2], \tag{A4}$$

or, if reaction is caused by the infiltration of fluids rather than a change in temperature,

$$\Delta G_r = n(X - X_{eq})[RT \ln\left(\frac{X}{1 - X}\right) - (1 - 2X)W_g], \tag{A5}$$

where n is the number of moles of fluid participating in the reaction, X_{eq} is the equilibrium composition of the fluid, X is the actual composition of the fluid, R is the gas constant, and W_g is Margules' parameter. For small variations in X,

$$\dot{n} \propto \exp[(X - X_{eq})^2]. \tag{A6}$$

2. Linear growth rates when these are controlled by the interface reaction (Christian, 1975, p. 479):

$$\dot{g} = A_g \exp(-\Delta G^*/kT)[1 - \exp(\Delta G_r/kT)]. \tag{A7}$$

A_g is a constant given approximately by $\delta_\beta\nu$, where δ_β is the thickness of a layer of ions.

For small amounts of overstepping this equation can be simplified to:

$$\dot{g} = A_g(\Delta G_r/kT) \exp(-\Delta G^*/kT). \tag{A8}$$

3. Diffusion-controlled growth rate for three-dimensional growth of an iso-

lated spherical nucleus at small degrees of supersaturation (Christian, 1975, p. 487, equations 54:8–54:12):

$$\frac{\partial r}{\partial t} = \dot{g} = \frac{\alpha_j}{2}\left(\frac{D}{t}\right)^{1/2}, \tag{A9}$$

where α_j is a geometry-specific constant related to the dimensionless supersaturation $\bar{\alpha}$:

$$\bar{\alpha} = \frac{c^m - c^\alpha}{c^\beta - c^\alpha}, \tag{A10}$$

where c^m is the matrix concentration of the diffusing species, c^β is the concentration in the matrix at the growing surface, and c^α is the concentration in the growing grain.

For the geometry considered here,

$$\begin{aligned}\alpha_j &= (2\bar{\alpha})^{1/2} \\ &= \sqrt{[2(c^m - c^\alpha)/(c^\beta - c^\alpha)]}.\end{aligned} \tag{A11}$$

Appendix II

Values of the temperature overstep of an equilibrium boundary $(T - T_{eq})$ required for a significant rate of nucleation (taken as $10^7 s^{-1}m^{-3}$, or approximately 1 yr.$^{-1}m^{-3}$) calculated from Eq. (A1), Appendix I, assuming a value for the free energy of activation of $\ln(10^{-8})T$.

$\Delta S_{reaction}$	Equivalent interfacial free energy $(\gamma)^a$		
	0.06 J m^{-2}	0.6 J m^{-2}	
1250 kJ m^{-3}K$^{-1 b}$	4	120	$(T - T_{eq})$
250 kJ m^{-3}K$^{-1 c}$	20	600	

[a] For substitution into Eq. (A1), Appendix I, these values must be reduced to 0.01 and 0.1 J m^{-2} to take into account the destruction of old grain boundaries when nucleation takes place at a grain corner (Cahn, 1956), assuming there are no significant differences between the interfacial free energies of the various phases). The values given above are equivalent to crystal surface energies of 0.2 and 2 Jm^{-2}.

[b] Value corresponds approximately to that of the reaction muscovite + quartz → sillimanite + K-feldspar + H_2O.

[c] Value corresponds to that of the reaction kyanite → sillimanite.

Taking the logarithmic average of the values of overstepping in the table above suggests a range of overstepping of 10–50 K in most metamorphic reactions. It is suggested that the extreme high values calculated may never

188									J. Ridley and A.B. Thompson

be important in metamorphism—there will always be a more favorable arrangement of minerals such that the effective interfacial free energy is lowered.

References

Aagaard, P., and Helgeson, H.C. (1982) Thermodynamic and kinetic constraints of reaction rates among minerals and aqueous solutions. 1. Theoretical considerations. *Amer. J. Sci.* **282**, 237–285.

Althaus, E. (1966) Die Bildung von Pyrophyllit und Andalusit zwischen 2000 und 7000 bar H_2O-Druck. *Naturwiss.* **53**, 105–106.

Althaus, E., Karotke, E., Nitsch, K.H., and Winkler, H.G.F. (1970) An experimental re-examination of the upper stability limit of muscovite plus quartz. *Neues Jahrb. Mineral. Monatsh.* **1970**, 325–336.

Atherton, M.P. (1964). The garnet isograd in pelitic rocks and its relations to metamorphic facies. *Amer. Miner.* **49**, 1331–1349.

Atherton, M.P. (1965). Chemical significance of isograds, in *Controls of Metamorphism,* edited by W.S. Pitcher and G.W. Flinn, pp. 169–202. Oliver and Boyd, Edinburgh.

Atherton, M.P. (1976) Crystal growth models in metamorphic tectonites. *Phil. Trans. Roy. Soc. London A* **283**, 255–270.

Atherton, M.P., and Edmunds, W.M. (1966) An electron microprobe study of some zoned garnets from metamorphic rocks, *E.P.S.L.* **1**, 185–193.

Beach, A. (1976) The interrelations of fluid transport, deformation and geochemistry and heat flow in early Proterozoic shear zones in the Lewisian complex. *Phil. Trans. Roy. Soc. London A* **280**, 529–604.

Becke, F. (1913) Ueber Mineralbestand und Struktur des Kristallinen Schiefer. *Denkschr. Akad. Wiss. Wien* **75**, 1–96.

Birch, F., and LeComte, P. (1960) Temperature-pressure plane for albite composition. *Amer. J. Sci.* **258**, 209–217.

Bloxam, T.W. (1960) Jadeite-rocks and glaucophane-schists from Angel Island, San Francisco Bay, California. *Amer. J. Sci.* **258**, 555–573.

Boettcher, A.L., and Wyllie, P.J. (1968) Jadeite stability measured in the presence of silicate liquids in the system $NaAlSiO_4$-SiO_2-H_2O. *Geochim. Cosmochim. Acta* **32**, 999–1012.

Boyd, F.R., and England, J.L. (1959) Pyrope. *Carnegie Inst. Washington Yearbook* **58**, 83–87.

Brace, W.F., and Walsh, J.B. (1962) Some direct measurements of the surface energy of quartz and orthoclase. *Amer. Miner.* **47**, 1111–1122.

Brady, J.B. (1979) Intergranular diffusion in quartz-periclase reaction couples. *Carnegie Inst. Washington Yearbook* **78**, 577–581.

Burnham, C.W., Holloway, J.R., and Davis, N.F. (1969) Thermodynamic properties of water to 1000° and 10,000 bars. *Geol. Soc. Amer. Spec. Paper* **132**, 96.

Cahn, J.W. (1956) The kinetics of grain boundary nucleated reactions. *Acta Metall.* **4**, 449–459.

Carlson, W.D., and Rosenfeld, J.C. (1981) Optical determination of topotactic aragonite-calcite growth kinetics: Metamorphic implications. *J. Geol.* **89**, 615–638.

Carmichael, D.M. (1969) On the mechanisms of prograde metamorphic reactions in quartz-bearing pelitic rocks. *Contrib. Mineral. Petrol.* **20**, 244–267.

Carmichael, D.M. (1979) Some implications of metamorphic reaction mechanisms for geothermobarometry based on solid-solution equilibria (abstract). *Geol. Soc. Amer. Abstr. Prog.* **11**, 398.

Carpenter, M.A., and Putnis, A. (1985) Cation order and disorder during crystal growth: Some implications for natural mineral assemblages, in *Metamorphic Reactions, Kinetics, Textures, and Deformation,* edited by A.B. Thompson and D.C. Rubie, pp. 1–26. Springer-Verlag, New York.

Chatterjee, N.D. (1970) Synthesis and upper stability of paragonite. *Contrib. Mineral. Petrol.* **27**, 244–257.

Chatterjee, N.D., and Johannes, W. (1974) Thermal stability and standard thermodynamic properties of synthetic 2 M_1-muscovite $KAl_2[AlSi_3O_{10}(OH)_2]$. *Contrib. Mineral. Petrol.* **48**, 89–114.

Chinner, G.A. (1961) The origin of sillimanite in Glen Clova, Angus. *J. Petrol.* **2**, 312–323.

Chinner, G.A. (1967) Chloritoid and the isochemical character of Barrow's zones. *J. Petrol.* **7**, 268–282.

Chinner, G.A., and Heseltine, F.J. (1979) The Grampide andalusite/kyanite isograd. *Scott. J. Geol.* **15**, 117–127.

Christian, J.W. (1975) *The Theory of Transformation in Metals and Alloys* (Second edition). *Part 1. Equilibrium and General Kinetic Theory.* Pergamon Press, Oxford.

Compagnoni, R., and Maffeo, B. (1973) Jadeite-bearing metagranitoids l.s and related rocks in the Monte Mucrone Area (Sesiz-Lanzo Zone, Western Italian Alps). *Schweiz. Min. Pet. Mitt.* **53**, 355–378.

Dachille, F., and Roy, R. (1964) Effectiveness of shearing stress in accelerating solid phase reactions at low temperatures and high pressures. *J. Geol.* **72**, 243–247.

Ehrlich, R., Vogel, T.A., Wemberg, B., Kamilli, D., Byerly, G., and Richter, H. (1972) Textural variation in petrogenetic analysis. *Geol. Soc. Amer. Bull.* **83**, 665–676.

Eskola, P. (1939) in *Die Entstehung der Gesteine,* by T.F.W. Barth, C.W. Correns, and P. Eskola. Springer Verlag, Berlin.

Etheridge, M.A., Wall, V.J., and Vernon, R.H. (1983) The role of the fluid phase during regional metamorphism and deformation. *J. Metam. Geol.* **1**, 205–226.

Eugster, H.P., and Yoder, H.S. (1954) Paragonite. *Carnegie Inst. Washington Yearbook* **53**, 111–114.

Evans, B.W. (1965) Application of a reaction rate method to the breakdown equilibria of muscovite and muscovite plus quartz. *Amer. J. Sci.* **263**, 647–667.

Ferry, J.M. (1983) Regional metamorphism of the Vassalboro Formation, south-central Maine, USA: A case study of the role of fluid in metamorphic petrogenesis. *J. Geol. Soc. London* **140**, 551–576.

Fisher, G.W. (1970) The application of ionic equilibria to metamorphic differentiation: An example. *Contrib. Mineral. Petrol.* **29**, 91–103.

Fisher, G.W. (1978) Rate laws in metamorphism. *Geochim. Cosmochim. Acta* **42**, 1035–1050.

Foster, C.T. (1981) A thermodynamic model of mineral segregations in the lower sillimanite zone near Rangeley, Maine. *Amer. Mineral.* **66**, 260–277.

Foster, C.T. (1983) Thermodynamic models of biotite pseudomorphs after staurolite. *Amer. Mineral.* **68**, 389–397.

Fowler-Billings, K. (1949) Geology of the Monadnock region of New Hampshire, *Geol. Soc. Amer. Bull.* **60**, 1249–1280.

Fox, J.S. (1975) Three-dimensional isograds from the Lukmanier Pass, Switzerland, and their tectonic significance. *Geol. Mag.* **112,** 547–564.

Freer, R. (1981) Diffusion silicate in minerals and glasses: A data digest and guide to the literature. *Contrib. Mineral. Petrol.* **76,** 440–454.

Fyfe, W.S. (1960) Hydrothermal synthesis and determination of equilibrium between minerals in the subliquidus region. *J. Geol.* **68,** 553–556.

Fyfe, W.S., Turner, F.J., and Verhoogen, J. (1958) Metamorphic reactions and metamorphic facies. *Geol. Soc. Amer. Mem.* **73,** 1–259.

Graham, C.M., Greig, K.M., Sheppard, S.M.F., and Turi, G. (1983) Genesis and mobility of the H_2O-CO_2 fluid phase during regional greenschist and epidote amphibolite facies metamorphism: A petrological and stable isotope study in the Scottish Dalradian. *J. Geol. Soc. London* **140,** 577–600.

Greenwood, H.J. (1963) The synthesis and stability of anthophyllite. *J. Petrol.* **4,** 317–351.

Guidotti, C.V. (1970a) The mineralogy and petrology of the transition from the lower to upper sillimanite zone in the Oquossoc area, Maine. *J. Petrol.* **11,** 277–336.

Guidotti, C.V. (1970b) Metamorphic petrology, mineralogy and polymetamorphism in a portion of N.W. Maine. New England International Geological Congress, Field Guide.

Guidotti, C.V. (1974) Transition from staurolite to sillimanite zone, Rangeley Quadrangle, Maine. *Geol. Soc. Amer. Bull.* **85,** 475–490.

Haas, H., and Holdaway, M.J. (1973) Equilibria in the system Al_2O_3-SiO_2-H_2O involving the stability limits of pyrophyllite, and thermodynamic data of pyrophyllite. *Amer. J. Sci.* **273,** 449–464.

Heinrich, C.A. (1982) Kyanite-eclogite to amphibolite facies evolution of hydrous mafic and pelitic rocks, Adula Nappe, Central Alps. *Contrib. Mineral. Petrol.* **81,** 30–38.

Helgeson, H.C., Murphy, W.M., and Aagaard, P. (1984) Thermodynamic and kinetic constraints on reaction rates among minerals and aqueous solutions. II. Rate constants, effective surface area, and the hydrolysis of feldspar, *Geochim. Cosmochim. Acta* **48,** 2405–2432.

Holland, T.J.B., and Richardson, S.W. (1979) Amphibole zonation in metabasites as a guide to the evolution of metamorphic conditions. *Contrib. Mineral. Petrol.* **70,** 143–148.

Hollister, L.S. (1969) Metastable paragenetic sequence of andalusite, kyanite, and sillimanite, Kwoiek area, British Columbia. *Amer. J. Sci.* **267,** 352–370.

Hollister, L.S. (1977) The reaction forming cordierite from garnet, the Khtada lake metamorphic complex, British Columbia. *Can. Mineral.* **15,** 217–229.

Johannes, W., Bell, P.W., Mao, H.K., Boettcher, A.L., Chipman, D.W., Hays, J.F., Newton, R.C., and Seifert, F. (1971) An interlaboratory comparison of piston-cylinder pressure calibration using the albite breakdown reaction. *Contrib. Mineral. Petrol.* **32,** 24–38.

Jones, K.A., and Galway, A.K. (1964) Study of possible factors concerning garnet formation in the rocks from Ardara, Co Donegal. *Geol. Mag.* **101,** 76–93.

Jones, K.A., and Galway, A.K. (1966) Size distribution, composition and growth kinetics of garnet crystals in some metamorphic rocks from the West of Ireland, *J. Geol. Soc. London* **122,** 29–44.

Kerrick, D.M. (1968) Experiments on the upper stability limit of pyrophyllite at 1.8 and 3.9 kbar water pressure. *Amer. J. Sci.* **266,** 204–214.

Kirkpatrick, R.J., Klem, L., Uhlmann, D.R., Hays, J.F. (1979) Rates and processes of crystal growth in the system anorthite-albite. *J. Geophys. Res.* **84,** 3671–3676.

Knipe, R.J. (1981) The interaction of deformation and metamorphism in slates. *Tectonophysics* **78**, 249–272.

Kretz, R. (1966) Grain size distributions for certain metamorphic minerals in relation to nucleation and growth. *J. Geol.* **74**, 147–173.

Kretz, R. (1973) Kinetics of the crystallization of garnet at two localities near Yellowknife. *Can. Mineral.* **12**, 1–20.

Kretz, R. (1974) Some models for the rate of crystallization of garnet in metamorphic rocks. *Lithos* **7**, 123–131.

Kridelbaugh, S.J. (1973) The kinetics of the reaction calcite + quartz = wollastonite + carbon dioxide at elevated temperatures and pressures. *Amer. J. Sci.* **273**, 757–777.

Kwak, T.A.P. (1974) Natural staurolite breakdown reactions at moderate to high pressures. *Contrib. Mineral. Petrol.* **44**, 57–80.

Lacy, E.D. (1965) Factors in the study of metamorphic reaction rates, in *Controls of Metamorphism*, edited by W.S. Pitcher and G.W. Flinn, pp. 140–154. Oliver and Boyd, Edinburgh.

Lasaga, A.C., Richardson, S.M., and Holland, H.D. (1977) The mathematics of cation diffusion and exchange between silicate minerals during retrograde metamorphism, in *Energetics of Geological Processes*, edited by S.K. Saxena and S. Bhattacharji, pp. 353–388. Springer-Verlag, New York.

Loomis, T.P. (1972) Coexisting aluminium silicate polymorphs in contact metamorphic aureols. *Amer. J. Sci.* **272**, 933–945.

Loomis, T.P. (1975) Reaction zoning of garnet. *Contrib. Mineral Petrol.* **52**, 285–305.

Loomis, T.P. (1976). Irreversible reactions in high-grade metapelitic rocks. *J. Petrol.* **17**, 559–588.

Loomis, T.P. (1979) A natural example of a metastable reaction involving garnet and sillimanite. *J. Petrol.* **20**, 271–292.

Loomis, T.P. (1982) Numerical simulation of the disequilibrium growth of garnet in chlorite-bearing aluminous pelitic rocks. *Can. Mineral.* **20**, 411–423.

Loomis, T.P. (1983) Compositional zoning of crystals: A record of growth and reaction history, in *Kinetics and Equilibrium in Mineral Reactions*, edited by S.K. Saxena, pp. 1–61. Springer Verlag, New York.

Loomis, T.P., and Nimick, F.B. (1982) Equilibrium in Mn-Fe-Mg aluminous pelitic compositions and the equilibrium growth of garnet. *Can. Mineral.* **20**, 393–410.

Matsushima, S., Kennedy, G.C., Akella, J., and Haygarth, J. (1967) A study of equilibrium relations in the systems Al_2O_3-SiO_2-H_2O and Al_2O_3-H_2O. *Amer. J. Sci.* **265**, 28–44.

Matthews, A., and Goldsmith, J.R. (1984) The influence of metastability on reaction kinetics involving zoisite formation from anorthite at elevated pressures and temperatures. *Amer. Mineral.* **69**, 848–857.

McLean, D. (1965) The science of metamorphism in metals, in *Controls of Metamorphism*, edited by W.S. Pitcher and G.W. Flinn, pp. 103–118. Oliver and Boyd, Edinburgh.

Meyer, J. (1983) The development of the high-pressure metamorphism in the Allalin Metagabbro (Switzerland). *Terra Cognita* **3**, 187.

Naggar, M.H., and Atherton, M.P. (1970) The composition and metamorphic history of some aluminium silicate-bearing rocks from the aureoles of the Donegal Granite. *J. Petrol.* **11**, 549–589.

Newton, R.C., and Smith, J.V. (1967) Investigations concerning the breakdown of albite at depth in the earth. *J. Geol.* **75**, 268–286.

Pitcher, W.S. (1965) The aluminium silicate polymorphs, in *Control of Metamor-*

phism, edited by W.S. Pitcher, and G.W. Flinn, pp. 329–341. Oliver and Boyd, Edinburgh.

Porter, D.A., and Easterling, K.E. (1981) *Phase Transformations in Metals and Alloys.* Van Nostrand Reinhold Company, New York.

Powell, R. (1978) *Equilibrium Thermodynamics in Petrology: An Introduction.* Harper and Row, London, 284 pp.

Putnis, A., and McConnell, J.D.C. (1980) *Principles of Mineral Behaviour.* Blackwell Scientific Publications, Oxford. 257 pp.

Råheim, A., and Green, D.H. (1974) Experimental determination of the temperature and pressure dependence of the Fe-Mg partition coefficient for coexisting garnet and clinopyroxene. *Contrib. Mineral. Petrol.* **48,** 179–203.

Ramberg, H. (1952) *The Origin of Metamorphic and Metasomatic Rocks.* University of Chicago Press, Chicago. 317 pp.

Rast, N. (1965) Nucleation and growth of metamorphic minerals, in *Controls of Metamorphism,* edited by W.S. Pitcher and G.W. Flinn, pp. 73–102. Oliver and Boyd, Edinburgh.

Richardson, S.M. (1974) Cation exchange reactions and metamorphism of high-grade pelites in central Massachusetts (abstract). *Geol. Soc. Amer. Progr. Abstr. Progs.* **6,** 1059.

Ridley, J. (1985) The effect of reaction enthalpy on the progress of a metamorphic reaction, in *Metamorphic Reactions, Kinetics, Textures and Deformation* edited by A.B. Thompson and D.C. Rubie, pp. 80–97. Springer-Verlag, New York.

Ridley, J., and Dixon, J.E. (1984) Reaction pathways during the progressive deformation of a blueschist metabasite: The role of chemical disequilibrium and restricted range equilibrium. *J. Metam. Geol.* **2,** 115–128.

Robie, R.A., Hemingway, B.S., and Fisher, J.R. (1978). Thermodynamic properties of minerals and related substances at 298.15 K and 1 bar (10^5 pascals) pressure and at higher temperature. *U.S. Geol. Surv. Bull.* **1452.** 456 pp.

Robinson, D. (1971) The inhibiting effect of organic carbon in contact metamorphic recrystallization of limestone. *Contrib. Mineral. Petrol.* **32,** 245–250.

Rosenfeld, J.L. (1969) Stress effects around quartz inclusions in almandine and the piezothermometry of coexisting aluminum silicates. *Amer. J. Sci.* **267,** 317–351.

Roy, R., and Osborn, E.F. (1954) The system Al_2O_3-SiO_2-H_2O. *Amer. Mineral.* **39,** 853–885.

Rubie, D.C., and Thompson, A.B. (1985) Kinetics of metamorphic reactions at elevated temperatures and pressures: An appraisal of available experimental data, in *Metamorphic Reactions, Kinetics, Textures, and Deformations,* edited by A.B. Thompson and D.C. Rubie, pp. 27–79. Springer-Verlag, New York.

Rumble, D., Ferry, J.M., Hoering, T.C., and Boucot, A.J. (1982) Fluid flow during metamorphism at the Beaver Brook fossil locality, New Hampshire. *Amer. J. Sci.* **282,** 886–919.

Rutter, E.H., and Brodie, K.H. (1985) The permeation of water into hydration shear zones, in *Metamorphic Reactions: Kinetics, Textures, and Deformation,* edited by A.B. Thompson and D.C. Rubie, pp. 242–249. Springer-Verlag, New York.

Sander, B. (1930) *Einführung in die Gefügekunde der geologischen Körper.* Springer-Verlag, Berlin.

Spear, F.S., and Selverstone, J. (1983) Quantitative *P–T* paths from zoned minerals, theory and tectonic applications. *Contrib. Mineral. Petrol.* **83,** 348–357.

Spry, A. (1969) *Metamorphic Textures,* pp. 350. Pergamon Press, Oxford.

Strens, R.G. (1968) Stability of Al_2SiO_5 solid solutions. *Mineral. Mag.* **36,** 839–849.

Thompson, A.B. (1970) A note on the kaolinite-pyrophyllite equilibrium. *Amer. J. Sci.* **268**, 454–458.

Thompson, A.B. (1976) Mineral reactions in pelitic rocks. *Amer. J. Sci.* **276**, 401–454.

Thompson, A.B. (1983) Fluid-absent metamorphism. *J. Geol. Soc. London* **140**, 533–549.

Thompson, A.B., and England, P.C. (1984) Pressure–temperature–time paths of regional metamorphism. II. Their inference and interpretation using mineral assemblages in metamorphic rocks. *J. Petrol.* **25**, 929–955.

Thompson, A.B., Tracy, R.J., Lyttle, P., and Thompson, J.B. (1977) Prograde reaction histories deduced from compositional zonation and mineral inclusions in garnet from the Gassetts schist, Vermont. *Amer. J. Sci.* **277**, 1152–1167.

Thompson, J.B., Jr. (1955) The thermodynamic basis for the mineral facies concept. *Amer. J. Sci.* **253**, 65–103.

Tracy, R.J., and McLellan, E.L. (1985) A natural example of the kinetic controls of compositional and textural equilibration, in *Metamorphism Reactions: Kinetics, Textures, and Deformation,* edited by A.B. Thompson and D.C. Rubie, pp. 48–137. Springer-Verlag, New York.

Tracy, R.J., and Richardson, S.M. (1978) The reaction forming cordierite from garnet, the Khtada Lake metamorphic complex, British Columbia: Discussion. *Can. Mineral.* **16**, 277–279.

Tracy, R.J., and Robinson, P. (1980) Evolution of metamorphic belts: Information from detailed petrologic studies, in *The Caledonides in the U.S.A.,* vol. 2, edited by D.R. Wones, pp. 189–195. Virginia Polytechnic Inst. and State Univ. Memoir.

Tracy, R.J., Robinson, P., and Thompson, A.B. (1976) Garnet compositions and zoning in the determination of temperature and pressure of metamorphism. *Amer. Mineral.* **61**, 762–775.

Vernon, R.H. (1976) *Metamorphic Processes,* pp. 247. George Allen and Unwin Ltd., London.

Vernon, R.H. (1978) Pseudomorphous replacement of cordierite by symplectite intergrowths of andalusite, biotite and quartz. *Lithos* **11**, 283–290.

Walther, J.V., and Wood, B.J. (1983) Mechanistic controls on prograde metamorphism (abstract). *Geol. Soc. Amer. Abstr. Progs.* **15**, 417.

Walther, J.V., and Wood, B.J. (1984) Rate and mechanism in prograde metamorphism. *Contrib. Mineral. Petrol.* **88**, 246–259.

White, S. (1975) Tectonic deformation and recrystallization of oligoclase. *Contrib. Mineral. Petrol.* **50**, 287–304.

Wood, B.J., and Walther, J.V. (1983) Rates of hydrothermal reactions. *Science* **222**, 413–415.

Yardley, B.W.D. (1977) The nature and significance of the mechanism of sillimanite growth in the Connemara schists, Ireland. *Contrib. Mineral. Petrol.* **65**, 53–58.

Yardley, B.W.D. (1981) Effect of cooling on the water content and mechanical behaviour of metamorphosed rocks. *Geology* **9**, 405–408.

Yardley, B.W.D., Leake, B.E., and Farrow, C.M. (1980) The metamorphism of Fe-rich pelites from Connemara, Ireland. *J. Petrol.* **21**, 365–399.

Yoder, H.S., and Eugster, H.P. (1955) Synthetic and natural muscovites. *Geochim. Cosmochim. Acta* **8**, 225–280.

Chapter 8
Mineral–Fluid Reaction Rates

J.V. Walther and B.J. Wood

Introduction

The nature and extent of interaction between a metamorphic rock and externally derived fluid depend on a number of physical and chemical variables.
The compositions of rock and fluid and the initial degree of disequilibrium
between them is obviously of prime importance. Second, the rates of fluid–
mineral reactions control the actual extents to which reequilibration may be
observed. These rates are, in turn, dependent on pressure and temperature
and on the initial and final grain size of the rock. Finally, the net fluid–rock
ratio determines the final equilibrium mineral assemblages that would be
observed if infinite time were allowed for reaction. Petrographic and isotopic
examinations of rocks that have been through the metamorphic "mill"
sometimes enable estimates of overall fluid–rock ratios to be made (e.g.,
Rumble *et al.,* 1982). These calculations require the assumption of equilibrium, however, so that in the event of substantial disequilibrium, the actual
fluid–rock ratio is underestimated. Several studies have demonstrated that
limestones (e.g., Rumble *et al.,* 1982; Graham *et al.,* 1983) and pelites
(Ferry, 1984) have equilibrated with large amounts (1 : 1 or more by volume)
of externally derived fluid during prograde metamorphism. These observations lead to important questions about the overall rate of the metamorphic
process, the origins of the fluids, and the nature of fluid flow at deep crustal
levels (Wood and Walther, this volume). Since they are calculated by assuming equilibrium and represent minimum estimates of fluid added to the rocks,
they also lead us to question whether the actual amount of fluid flow was
much larger than 1 : 1 by volume and whether insufficient time was available
for complete equilibration. In order to address this issue it is necessary to
determine the rates of mineral–fluid reaction under metamorphic conditions.
That is the purpose of this paper.

Most of the available information on mineral–fluid reaction rates comes
from studies at 25°C. The experiments are generally performed by crushing

the mineral sample, measuring its surface area by gas adsorption, and then adding a known amount of it to a solution of controlled pH. The rate of dissolution is determined by analysis of the solution as a function of time. This method or variants of it have been employed in studies at temperatures up to about 300°C (e.g., Lagache, 1965; Rimstidt and Barnes, 1980) but have not, as yet, been extended into the higher temperature metamorphic regime. Mineral–fluid reaction rates under metamorphic conditions, although little studied, have been estimated by Wood and Walther (1983) from phase equilibrium experiments in which a single crystal of one of the reactants was used to monitor reaction direction. We will commence by reviewing the available mineral dissolution data and continue with a discussion of the temperature dependence of dissolution rates before putting them in the context of heterogeneous reaction under metamorphic conditions.

Mineral Dissolution at 25°C

Dissolution of silicate minerals can be viewed as the combination of three successive steps. First, if alkali or alkaline earth elements are present initial hydrolysis exchanges these elements on the mineral surface for hydronium ions in solution. This exchange occurs in a number of minutes at 25°C and generally penetrates about one unit cell deep (Tamm, 1930; Nash and Marshall, 1956; Garrels and Howard, 1959). After the rapid steady-state surface exchange, silicate minerals are observed to dissolve at a rate that is parabolic:

$$\frac{dm}{dt} = \tfrac{1}{2}k't^{-1/2} \qquad (\text{mole cm}^{-2}\ \text{s}^{-1}), \tag{1}$$

where k' is a constant at constant pressure, temperature, and solution composition. This parabolic behavior has been attributed to initial rapid dissolution of fine particles that adhere to the surfaces of coarser grains in crushed samples. Acid pretreatment of samples to remove the fine particles results in a close approach to linear dissolution kinetics (Holdren and Berner, 1979).

Chou and Wollast (1984) have shown that initial dissolution of albite is incongruent with respect to both Na and Al relative to Si. On the basis of their experiments, a depleted layer whose average thickness is roughly 30 Å forms on the feldspar surface. The composition of the layer is pH dependent. From the rapid change in calculated composition of this layer with changing pH, Chou and Wollast argue that the depleted layer may be extremely important in early dissolution processes. In most cases, however, after a few tens of days or less at 25°C the reaction rate becomes steady state and linear:

$$\frac{dm}{dt} = k \qquad (\text{mole cm}^{-2}\ \text{s}^{-1}). \tag{2}$$

This third type of dissolution behavior that dominates on the time scale of geologic processes has been observed for feldspar (Holdren and Berner, 1979), quartz (Rimstidt and Barnes, 1980), phyllosilicates (Lin and Clemency, 1981a, 1981b), and chain silicates (Schott et al., 1981). Linear reaction rates imply that the rate-determining step for silicate dissolution involves reaction at the mineral–fluid interface. The lack of any significant observed stirring effect on dissolution rate rules out the possibility that diffusion in the fluid is rate determining (Berner, 1978). All these experiments were performed far from equilibrium so that the measured rate of reaction is very close to the rate of the forward dissolution reaction since negligible back (precipitation) reaction can occur under such circumstances. Dissolution rates vary widely from mineral to mineral, however, and they are markedly pH dependent (e.g., Chou and Wollast, 1985).

The simplest theoretical analysis of dissolution kinetics follows from transition state theory (Wynne-Jones and Eyring, 1935; Eyring, 1935), which has previously been discussed in the geological literature by Lasaga (1981) and Aagaard and Helgeson (1982). In this approach, the reactants (mineral and fluid) are assumed to be in equilibrium with a high-energy activated complex or combination of activated complexes. In alkali feldspars, these complexes might have formulae similar to hydrated or alkali-free charged feldspars (e.g., Helgeson et al., 1984). The rate-determining step involves irreversible decomposition of the activated complex to form product species. The reverse, growth reaction takes place via assumed equilibrium between the product species and the activated complex with a rate controlled by irreversible decomposition to form the mineral plus fluid. In each case, the rate of reaction depends on the free energy difference between the activated complex and the reactants. Thus, conceptually we have:

$$\text{mineral} + \text{fluid} \rightleftharpoons \text{activated complex} \rightarrow \text{product species} \qquad (\text{dissolution})$$

$$\frac{dm}{dt} \propto e^{-(G_{ac} - G_{min} - G_{fluid})/RT}$$

$$\text{product species} \rightleftharpoons \text{activated complex} \rightarrow \text{mineral} + \text{fluid} \qquad (\text{growth})$$

$$\frac{dm}{dt} \propto e^{-(G_{ac} - G_{prod})/RT}$$

The exponential factor arises by application of Boltzmann's law to the probability of the surface species having the necessary energy to form the activated complex, and G_i refers to the Gibbs free energy of the subscript species. Only at equilibrium can both of the above relations be strictly true. The relations imply that reactants and products are in equilibrium with the same activated complex, which in turn implies that the total reaction is at equilibrium. Two possibilities, therefore, exist. Either the activated complexes are not the same for growth and dissolution or it is not strictly valid to consider equilibrium between the activated complex and the reactants to be maintained during dissolution or between activated complex and produces during growth. As an approximation, however, we can assume similar complexes for dissolution and growth and argue that during dissolution reactants are not strictly in equilibrium with the activated complex but "near enough" to allow the above proportionality to be maintained. Similar arguments hold for growth rates. With this approximation, the ratio of dissolution to growth rates depends on the free energy of the overall reaction under the conditions of P, T, a_{H_2O}, etc., of concern. A simple transformation (Lasaga, 1981, p. 146) leads to a net observable rate of reaction (i.e., of dissolution minus growth) given by:

$$\frac{dm}{dt} = k(1 - e^{\Delta G_R/RT}) \qquad (\text{mole cm}^{-2}\ \text{s}^{-1}), \qquad (3)$$

where ΔG_R is the Gibbs free energy of the overall reaction ($G_{prod} - G_{min} - G_{fluid}$) under the conditions of interest and k is the rate constant for the forward, dissolution, reaction.

In Fig. 1, Eq. (3) is plotted as a function of $\Delta G_R/RT$ and dm/dt. Inspection of Fig. 1 shows that if a mineral is dissolved in a fluid with which it is far from equilibrium so that ΔG_R is large and negative, Eq. (3) becomes:

$$\frac{dm}{dt} = k \qquad (\Delta G_R/RT < -6), \qquad (4)$$

while under near equilibrium conditions we have:

$$e^{\Delta G_R/RT} = 1 - \frac{\Delta G_R}{RT} \qquad (\Delta G_R/RT > -0.4),$$

and

$$\frac{dm}{dt} = -k \frac{\Delta G_R}{RT}. \qquad (5)$$

Eq. (4) shows that the rate of surface reaction will yield the observed linear kinetics under dissolution conditions far from equilibrium. The near-equilib-

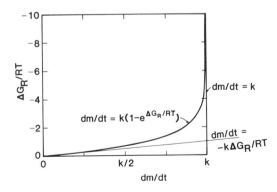

Fig. 1. General form of the rate of surface reaction dm/dt as a function of free energy driving force $\Delta G_R/RT$. Units of dm/dt are k mole cm^{-2} s^{-1}.

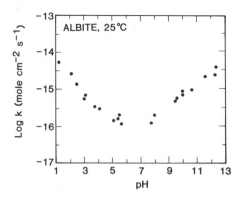

Fig. 2. Logarithm of the dissolution constant, k, for albite in mole cm^{-2} s^{-1} plotted against pH at 25°C. Data from Chou and Wollast (1985).

rium conditions relevant to Eq. (5) are, as will be shown below, more appropriate to metamorphism. The value of k is, however, identical in both Eq. (4) and (5) and is the rate constant for the forward, dissolution reaction.

A number of general observations may be made about the values of k for dissolution reactions. At any particular pH the observed values of k (mole cm^{-2} s^{-1}) for silicate minerals vary by about 5 orders of magnitude at 25°C depending on the mineral involved. Additionally, the rate constant for any particular silicate mineral is pH dependent. Shown in Fig. 2 are experimentally measured values of k for albite dissolution at 25°C and 1 bar determined by Chou and Wollast (1985). As can be seen, the rate constant increases dramatically with decreasing pH at low pH and with increasing pH at high pH, while between pH 5 and 8 there is a low-rate region in which the pH dependence is small. Although it seems likely that most minerals exhibit similar dissolution behavior, with a near neutral region of pH independence, the best documented phenomenon is of decreasing rate constant with increasing pH in the acid region. In support of a general minimum in silicate dissolution rates under near-neutral conditions, it has recently been found that basaltic glass shows a similar trend to that of Fig. 2 (Mazer and Walther, 1985).

As stated above, observed silicate dissolution rates at 25°C vary by about 5 orders of magnitude depending on the mineral involved. These variations should, however, be understandable crystallochemically because the dissolution reactions are similar and all involve breaking bonds in silicon–oxygen tetrahedra. Furthermore, there is an often-discussed weathering sequence among minerals that is related to the igneous reaction series of Bowen. In general, it is found that the rates of weathering are olivine > pyroxene > biotite among mafic minerals and anorthite > albite > K-feldspar > quartz among felsic minerals (Goldich, 1938; Krauskopf, 1967, p. 105; Lasaga, 1984). Qualitatively, this weathering series reflects a relationship between k and the relative stability between minerals at earth surface conditions. It might seem reasonable to expect a correlation between k and Gibbs free energy per unit volume or gram atom of oxygen. No simple correlation exists. At 25°C stishovite has a Gibbs free energy of formation of more than 12 kcal mol^{-1} greater than quartz, yet quartz dissolves more rapidly than stishovite (R.M. Garrels, personal communication). Clearly, the concept of relative stability and reaction rates is limited. Within the context of transition state theory, k_0' depends primarily on the standard state entropy of activation (see below). The values of k_0' are directly proportional to the rate constant k at a fixed temperature (Eq. (10)). Therefore, we may expect to find a correlation between k and relative entropies per unit volume or gram atom of oxygen. Minerals with high entropy per unit volume (S^0/V^0) dissolve more rapidly at low temperatures than do those with low entropy per unit volume (Fig. 3).

Fig. 3 shows a plot of log k (gram atom oxygen cm^{-2} s^{-1}) versus S^0/V^0 for minerals for which 25°C dissolution data in the near-neutral region are available. (In order to facilitate comparison between minerals with different numbers of atoms in the formula unit, k has been normalized to a constant gram atom of oxygen basis.) With a couple of exceptions there appears to be a

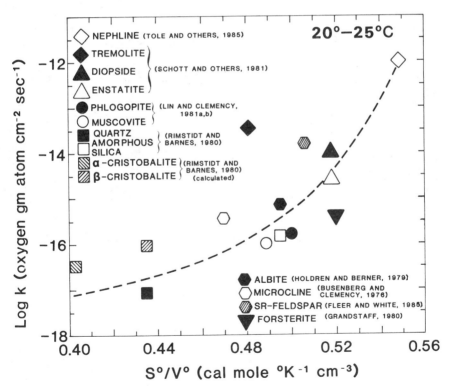

Fig. 3. Logarithm of the dissolution constant, k (oxygen gram atom cm^{-2} s^{-1}), at 20–25°C plotted against the 25°C entropy per unit volume, S^0/V^0, in cal mol^{-1} °C^{-1} cm^{-3}. The dissolution constants refer to near-neutral pH solutions.

reasonably good correlation between S^0/V^0 and log k at 25°C. The correlation is far from perfect, of course. Tremolite is predicted to have a rate constant 2.5 orders of magnitude lower than the measured value, and there is no obvious reason for the discrepancy. The forsterite k is predicted to be 1 order of magnitude greater than the measured value. The observed forsterite dissolution rate is particularly puzzling because it is lower than reported values for enstatite and diopside. Since beach sands composed of olivine and pyroxene generally show the opposite behavior, with olivine dissolving more readily than pyroxene, it is possible that the experimental data are in error.

What Fig. 3 shows, in the general context of transition state theory, is that different silicates produce different concentrations of the activated surface complex, the most unstable (greater entropy) producing the greatest abundance of activated complexes. If the activated complexes on the surfaces of the various mineral species have similar potential energies then their rates of breakdown should also be similar. Therefore, the more unstable minerals (higher S^0/V^0) should dissolve at higher rates than more stable minerals because they produce greater concentrations of activated complex per unit surface area.

Mineral Dissolution at High Temperature

In the context of transition state theory, the temperature dependence of the dissolution rate constant k is given by (Lasaga, 1981):

$$\frac{d\ln k}{dT} = \frac{1}{T} + \frac{\Delta H^{0\prime}}{RT^2}, \tag{6}$$

where $\Delta H^{0\prime}$ is the standard state enthalpy change of the reaction to produce the activated complex:

$$\text{mineral} + \text{fluid} \rightleftharpoons \text{activated complex},$$

if the heat capacity of activation ($\Delta C_P^{0\prime}$) is assumed to be close to zero. Eq. (6) may be compared with the Arrhenius form of the temperature dependencies of rate constants:

$$k = k_0 e^{-\Delta E_a/RT} \tag{7}$$

$$\frac{d\ln k}{dT} = \frac{\Delta E_a}{RT^2}. \tag{8}$$

In Eqs. (7) and (8) ΔE_a is the "energy of activation," a parameter which, in transition state theory, is closely related to the standard state enthalpy change of the reaction to produce the activated complex. In general, the rate constants increase with increasing temperature and the observed temperature effect depends on the height of the energy barrier to produce the activated complex.

Wood and Walther (1983) obtained rate constants for mineral–fluid reactions at high temperatures (up to 750°C) using phase equilibrium results from experiments in which a single crystal was used to monitor reaction direction. Under near-equilibrium conditions they used Eq. (5) in conjunction with measured weight changes of the single crystals and the assumption that the rate determining step involved reaction at the surface of the single crystal. It was found that, when normalized to a constant number of oxygen gram atoms, virtually all minerals from orthosilicates through tektosilicates dissolve (or grow) with essentially the same rate constant in the high-temperature regime. Furthermore, no significant dependence on pressure could be found. Wood and Walther therefore proposed a single rate equation for silicate dissolution and precipitation involving high-temperature aqueous and H_2O-CO_2 fluids. The equation:

$$\log k = \frac{-2900}{T} - 6.85 \qquad \text{(gram atom oxygen cm}^{-2} \text{ s}^{-1}), \qquad (9)$$

was assumed to be applicable to most silicate minerals under near-neutral conditions at temperatures to 750°C.

A Possible Compensation Relationship

Since mineral–fluid reaction rate constants are similar for most species at high temperatures but differ by 5 orders of magnitude at 25°C, they must converge as temperature is raised above 25°C. In practice this means that an Arrhenius plot of $\log k$ versus $1/T$ for each of the different minerals will produce a family of curves (virtually straight lines in this case) that are fanned out at low temperatures and that converge in the metamorphic temperature range (Fig. 5). Integration of Eq. (6) yields a result similar to the Arrhenius Eq. (7):

$$k = k_0' T e^{-\Delta H^{0'}/RT}, \qquad (10)$$

where k_0' as shown by Helgeson $et\ al.$ (1984) is independent of temperature and related to k_0 by:

$$k_0' = \frac{k_0}{T \exp(1)}. \qquad (11)$$

Convergence of the $\log k$ versus $1/T$ curves for different minerals means that there is a compensation relationship between the k_0' and $\Delta H^{0'}$ terms of Eq. (10) (or k_0 and ΔE_a terms of Eq. (7)). An exactly analogous relationship has been observed between the diffusion coefficients of alkali ions in alkali silicate glasses (Winchell, 1969). In practice, such compensation relationships mean that the preexponential terms in Eqs. (7) and (10) tend to increase as the enthalpy of activation $\Delta H^{0'}$ increases. Therefore, ΔE_a or $\Delta H^{0'}$ should be inversely proportional to $\log k$ at any temperature below that at which the curves all cross. As a test of the proposed compensation relationship we have plotted in Fig. 4 some available data on activation energy ΔE_a for low temperature dissolution against the observed 25°C values of log k. With the exception of the datum for nepheline dissolution, there does seem to be an inverse correlation between ΔE_a and log k. The solid line in Fig. 4 gives the following compensation relationship between ΔE_a and log k at 25°C:

$$\Delta E_a = 50,000 + 5300 \log k_0 \qquad (\text{cal mol}^{-1}), \tag{12}$$

which, in terms of transition state theory, transforms to:

$$\Delta H^{0\prime} = 64,800 + 5300 \log k_0^\prime. \tag{13}$$

The compensating relationship between $k_0^\prime$ and $\Delta H^{0\prime}$ may be qualitatively explained in the following way. In transition state theory $k_0^\prime$ depends primarily on the standard state entropy of activation $\Delta S^{0\prime}$ (Lasaga, 1981, p. 143). In passing from the stable surface state to the activated complex atoms at the crystal surface become less tightly bound to the crystal structure and increase in enthalpy and entropy. The greater the crystallochemical differences between stable and activated states, the larger the enthalpy $\Delta H^{0\prime}$ and entropy $\Delta S^{0\prime}$ changes associated with activation. Conversely, if stable and activated states are crystallochemically similar, the enthalpy and entropy changes associated with activation will be small. Therefore, one would anticipate an approximately positive correlation between $\Delta S^{0\prime}$ and $\Delta H^{0\prime}$ and hence between log $k_0^\prime$ and $\Delta H^{0\prime}$. This is the correlation shown in Fig. 4 and represented by Eq. (12) and (13). If the 25°C rate constant is known (or estimated by the correlation shown in Fig. 3) $\Delta H^{0\prime}$ and $k_0^\prime$ can be calculated by solving Eq. (10) and (13). Eq. (10) can then be used to calculate the temperature dependence of the rate constant.

Fig. 5 shows experimentally derived values of log k in oxygen gram atom $\text{cm}^{-2} \text{ s}^{-1}$ versus the reciprocal of temperature. The experimental values

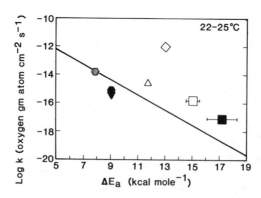

Fig. 4. Logarithm of the dissolution constant, k (oxygen gram atom $\text{cm}^{-2} \text{ s}^{-1}$), plotted against the activation energy ΔE_a in kcal mol^{-1}. Key to symbols is given in Fig. 3. The solid line gives the compensation relationship described in the text and shown in Fig. 5.

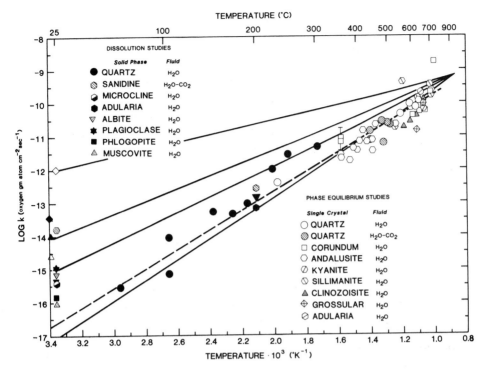

Fig. 5. Logarithm of the dissolution constant, k (oxygen gram atom cm^{-2} s^{-1}), plotted against the reciprocal of temperature ($^{\circ}$K^{-1}). The data are for near-neutral pH solutions. Symbols not keyed on figure are from Fig. 3. The dashed line gives the rate constant proposed by Wood and Walther (1983). The solid lines give the rate constants for the compensation law proposed in the text constructed for log k of -12, -14, -15, -17 at 25°C with the aid of Eqs. (10) and (13).

shown are identical to the data compiled by Wood and Walther (1983) with the addition of those shown in Fig. 3. The dashed line is the rate constant dependence on temperature proposed by Wood and Walther. The solid lines were constructed using Eqs. (10) and (13) for log k of -17, -15, -14, and -12 oxygen gram atom cm^{-2} s^{-1} at 25°C. Note that because of the proposed compensation between $\Delta H^{0\prime}$ and k_0', the computed values of log k converge with increasing temperature. If most minerals are characterized by log k between -17 and -14 oxygen gram atom cm^{-2} s^{-1} at 25°C then Fig. 5 shows that at 300°C they should all have log k values between roughly -12 to -11 oxygen gram atom cm^{-2} s^{-1}. At 500°C they should only vary by half an order of magnitude.

It appears that the conclusions of Wood and Walther (1983) concerning rates of mineral–fluid reaction at elevated temperatures are consistent with 25°C dissolution data for most minerals provided a compensation law is assumed to hold. Measured activation energies confirm the likelihood of compensation and support an essentially universal rate law in the metamorphic temperature range. Therefore, for the purposes of computing rates of mineral–fluid equilibration during prograde metamorphism we may use the approximate relationship (Wood and Walther, 1983):

$$\log k = \frac{-2900}{T} - 6.85 \qquad \text{(oxygen gram atoms cm}^{-2}\text{s}^{-1}).$$

It is now appropriate to consider the implications of these relationships for rates of fluid–rock equilibration and calculated fluid–rock ratios in prograde regimes. Much of the discussion presented here is based on the equations of Walther and Wood (1984), and the reader is referred to that paper for more detailed derivations.

Rates of Mineral–Fluid Equilibration

In terms of prograde metamorphic reactions, the extents of disequilibrium depend on the rates of temperature rise, surface reaction, and material transport and on whether or not the stable assemblage may nucleate. Leaving nucleation and transport aside for the moment, it may readily be shown that equilibration between product and reactant minerals should be rapid in the case of a closed system.

Consider a generalized dehydration or decarbonation reaction:

$$A + B = C + H_2O \text{ and/or } CO_2.$$

Under near-equilibrium conditions, the rate of removal and addition of material to reactant and product minerals is given by Eq. (5):

$$\frac{dm}{dt} = -k\frac{\Delta G_R}{RT} \qquad \text{(gram atom oxygen cm}^{-2}\text{s}^{-1}).$$

The free energy driving force is equal to the entropy change of reaction (about 20 cal/K per mole of fluid) multiplied by the temperature difference $(T - T_{\text{equilibrium}})$. A typical regional metamorphic event would have a rate of temperature rise of 10°C/m.y. or 3.2×10^{-13} °C s^{-1} so that we obtain:

$$\Delta G_R = 20 \times 3.2 \times 10^{-13} \times t \qquad \text{(cal mol}^{-1}),$$

where t is the time in seconds since the equilibrium boundary was crossed. The rate of transfer of material to and from products and reactants is in gram atoms oxygen cm^{-2} s^{-1}, a quantity that is more conveniently treated in terms of the rate of change of the radius of the crystal. As discussed by Walther and Wood (1984), this transformation is performed by dividing by the number of gram atoms of oxygen per cm^3 of crystal, about 0.09 for quartz:

$$\frac{dr}{dt} = -k\left(\frac{20 \times 3.2 \times 10^{-13} \times t}{RT \times 0.09}\right) \qquad \text{(cm s}^{-1}).$$

Substituting, as examples, values of k for 500°C (2.5×10^{-11}) and temperature of 773 K we obtain:

$$\frac{dr}{dt} = 1.157 \times 10^{-24} \, t \qquad (cm \ s^{-1}). \qquad (14)$$

Integrating and rearranging gives:

$$t = \left(\frac{r - r^0}{5.8 \times 10^{-25}}\right)^{1/2} \ seconds.$$

If we assume that dissolving and growing crystals are on the order of 0.1 cm radius, then the time for reaction to go to completion is 4×10^{11} seconds or 13,000 yr. Although this seems a long time, the reaction is actually completed while heating to only 0.13°C above the equilibrium boundary. Obviously, the rate of surface reaction can keep up with the rate of temperature rise in a regional metamorphic event.

Transport of material from sites of mineral dissolution to those of growth will occur by a combination of fluid flow and molecular diffusion through the fluid phase. Whether or not transport keeps up with surface reaction rate depends, therefore, on the amount of fluid passing through the rock since, under metamorphic conditions, diffusion through dry fluid-free grain boundaries is an extremely inefficient transport process (Walther and Wood, 1984).

The volume of fluid passing through any particular rock depends on how much fluid is being generated, the presence or absence of preferential flow pathways, such as fractures, and whether or not there is any recirculation. For reasons presented extensively elsewhere (Wood and Walther, this volume), fluid is assumed to flow out of the system only and not to be recirculated. A rock situated halfway down an 8-km column undergoing metamorphism at 10°C/m.y. will experience an average fluid flux of about 8.5×10^{-11} g cm^{-2} s^{-1} or 9.4×10^{-11} cm^3 cm^{-2} s^{-1} (Walther and Wood, 1984, p. 255). If all of this fluid flows along grain boundaries and none of it is preferentially channeled, then surface reaction rates will be slower than transport rates and the former will be the slowest step in the metamorphic process. One-way flow along grain boundaries toward the earth's surface yields, in limestone, fluid–rock ratios on the order of 2 : 1 by volume if carbonates occupy 5% of the metamorphic terrane (Wood and Walther, this volume). Although these values are roughly consistent with the petrographic and isotopic data of Rumble et al. (1982) and Graham et al. (1983), externally derived fluid from igneous intrusions or from dewatering of subducting lithosphere provides a potential for even higher fluid–rock ratios. It seems probable, therefore, that the high fluid–rock ratios in metacarbonates require average fluid fluxes on the order of 9.4×10^{-11} cm^3 cm^{-2} s^{-1} or greater during metamorphism. At the observed low values of X_{CO_2} (<0.15), transport of material through the fluid would be at least as rapid as surface reaction rate, and we therefore anticipate the latter to be rate limiting (Wood and Walther, this volume).

Let us now consider the rate of equilibration of a carbonate rock with infiltrating H_2O-rich fluid. Infiltration along grain boundaries (modeled as flow along tabular fractures) leads to a flux q given by:

$$q = \frac{d^3 \, l \, \tau}{12 \, \xi} \left(\frac{\delta P}{\delta Z}\right)_{\text{viscous}} \quad (\text{cm}^3 \, \text{cm}^{-2} \, \text{s}^{-1}). \tag{15}$$

In Eq. (15), q is the volume flux, d is the thickness of the fluid film flowing along grain boundaries, l is the length of grain boundary per cm^2 of rock surface, ξ is the fluid viscosity, $(\delta P/\delta Z)_{\text{viscous}}$ is the viscous pressure gradient, and τ is the tortuosity. Typical values of ξ and τ are 10^{-3} poise and 0.7, respectively, and the viscous pressure gradient is 1.9×10^3 dyne cm^{-3} (Walther and Orville, 1982; Walther and Wood, 1984). The length of grain boundary per unit area of rock surface l is roughly equal to $1/r$ where r is the average grain radius. We have, therefore, if q is 9.4×10^{-11} $\text{cm}^3 \, \text{cm}^{-2} \, \text{s}^{-1}$ and r is about 0.1 cm:

$$d = \left(\frac{12 \times 10^{-3} \times 9.4 \times 10^{-11}}{10 \times 0.7 \times 1900}\right)^{1/3} = 4.4 \times 10^{-6} \text{ cm.}$$

The average fluid velocity through the rock is:

$$v = \frac{q}{dl} = 2.1 \times 10^{-6} \text{ cm s}^{-1} = 67 \text{ cm/yr.}$$

If the fluid were initially 99 and 1 mol % of H_2O and CO_2 respectively, it would presumably stimulate decarbonation reactions in the rock. If we suppose that the equilibrium X_{CO_2} for a reaction occurring under these P–T conditions is 0.1 (cf. Rumble et al., 1982), it is of interest to ask how long it would take the rock to produce sufficient CO_2 to approach equilibrium. If all mineral reactants are pure, the free energy driving force for the reaction at any X_{CO_2} will be:

$$\Delta G_R = +RT \, \ln \left(\frac{X_{CO_2}}{X_{CO_2}^0}\right),$$

where $X_{CO_2}^0$ is the equilibrium value of 0.1. Substituting into Eq. (5) we have:

$$\frac{dm}{dt} = -k \, \ln \left(\frac{X_{CO_2}}{X_{CO_2}^0}\right).$$

Note that because Eq. (5) only holds near equilibrium, this calculation will underestimate the initial rate of reaction somewhat. Transforming dm/dt into dX_{CO_2}/dt by multiplying by the number of cm^3 per mole of CO_2 (~25 at high P and T), dividing by the number of cm^3 of fluid per cm^2 of surface area, $d/2$, and assuming that the reactive surface area is 20% of the whole, yields:

$$\frac{dX_{CO_2}}{dt} = \frac{-10k}{d} \, \ln \left(\frac{X_{CO_2}}{X_{CO_2}^0}\right).$$

Rearranging and integrating between X_{CO_2} of 0.01 and a near equilibrium value of 0.099 gives at 500°C:

$$\left\{ \ln \ln\left(\frac{X_{CO_2}}{X_{CO_2}^0}\right) + \ln\left(\frac{X_{CO_2}}{X_{CO_2}^0}\right) + \frac{\left[\ln\left(\frac{X_{CO_2}}{X_{CO_2}^0}\right)\right]^2}{2\cdot 2!} + \frac{\left[\ln\left(\frac{X_{CO_2}}{X_{CO_2}^0}\right)\right]^3}{3\cdot 3!} + \cdots \right\}_{0.01}^{0.099}$$

$$= \frac{-2.5 \times 10^{-10}}{d}\, t.$$

Using the value of grain boundary film thickness d calculated above (4.4 $\times$ 10^{-6} cm), t is found to be 110,000 s or just over 30 hr. Thus, the rate of surface reaction is sufficiently high to generate a near-equilibrium CO_2/H_2O ratio in the fluid within 30 hr. In this time, the fluid will have travelled only 0.23 cm. Therefore, for the case of fluid generated solely within the metamorphic pile a carbonate rock infiltrated by aqueous fluid will rapidly buffer its fluid composition. The situation is little different if there are large fluxes of externally derived fluid. For example, if there were ten times as much aqueous fluid passing through the rock, equivalent to an underlying dehydrating column 40 km thick, the fluid would have a film width of 9.5 $\times$ 10^{-6} cm and would equilibrate in 235,000 s. This situation, which might be appropriate for metamorphism in continent–continent or arc–continent collision zones, would still lead to equilibration of the fluid in a path length through the rock of 2.3 cm. Therefore, it appears that given nucleation of reactant and product minerals, carbonate rocks will buffer infiltrating fluid composition until all of their CO_2 is used up. The fluid–rock ratios exhibited by these rocks should be near-equilibrium values, and the rock should reach its final state without allowing unreacted fluid to pass. The only way to avoid these conclusions is if the fluid is channeled in major fractures such that the fluid will not contact the bulk of the carbonate unit.

Coupling of Surface Reaction, Transport, and Mineral Nucleation

Nucleation during prograde metamorphism is generally thought to occur heterogeneously at high energy sites such as grain boundaries and dislocations (Spry, 1969). In a way analogous to the free energy of overstepping (ΔG_R) required to cause net surface dissolution, a free energy driving force is needed to initiate nucleation. Obviously the nucleation barrier is generally crossed because new phases appear in a regular manner at isograds. Furthermore, the newly nucleated phases appear systematically in the zone of reaction while the disappearance of reactant phases is often irregular. Nucleation of any particular phase also occurs continuously throughout the metamorphic process and not just at specific isograds (e.g., Jones and Galwey, 1966; Kretz, 1966, 1973). These observations, coupled with the fact that, where observed, disequilibrium can be documented precisely because phases have nucleated but not completely grown or completely disappeared, leads us to suggest that nucleation is not generally rate limiting in metamorphism. Rather, it is more likely to be coupled to surface reaction and species transport.

Continuing nucleation of a phase after its isograde has been crossed implies that transport rates are not keeping up with surface reaction rates. The newly added product phase helps absorb the "excess capacity" generated by the surface reaction. Not only are the newly nucleated sites sinks for material but they will, in general, decrease the average distance between reactants and products and thus enhance diffusive transport because the shorter distance increases chemical potential gradients. Such a scenario of coupling surface reaction rate to transport and nucleation rate is consistent with the principle of entropy minimization in the thermodynamics of irreversible processes (e.g., Katchalsky and Curran, 1965, Chap. 14). Coupling of the rate of surface reaction to that of transport implies that, through increased nucleation and transport abilities, overall reaction rates should not depart significantly from the surface reaction rate in the presence of an aqueous fluid. It is also to be expected that the distance between reactants and products is a qualitative measure of transport ability. In the case of pseudomorphing one would surmise that transport ability was at a minimum. This is consistent with the common observation of pseudomorphing during retrograde metamorphism where the amounts of aqueous fluid involved are very low and the rates of material transport are correspondingly slow. Exceptions to general surface reaction control might occur where chemical potential gradients are induced by compositional differences rather than by overstepping of reaction temperature. Under these conditions, large chemical potential gradients could be maintained and reaction rates would be controlled by transport processes (Walther and Wood, 1984).

Conclusions

At 25°C, silicate dissolution rates are pH dependent and vary widely from one mineral to another. At high temperatures (300–700°C), however, the available data indicate that most minerals dissolve in aqueous and H_2O-CO_2 fluids with similar rate constants if their rates are normalized to a constant number of oxygen gram atoms. These observations suggest the presence of a compensation relationship between preexponential (k_0) and activation energy (ΔE_a) terms in the Arrhenius equation:

$$k = k_0 \, e^{-\Delta E_a/RT}$$

In our analysis of low-temperature dissolution data, we have found general confirmation of such a compensation relationship. We have also found a semiempirical correlation between 25°C dissolution rate constants and the entropy per unit volume ($S°/V°$) of the dissolving mineral. These observations enable approximate calculation of k for mineral–fluid reactions in the 25–750°C temperature range at near-neutral pH. At temperatures above 300°C the relationship is summarized by the equation of Wood and Walther (1983):

$$\log k = \frac{-2900}{T} - 6.85 \qquad (\text{gram atom oxygen cm}^{-2}\ \text{s}^{-1}).$$

Use of this equation shows that isograd reactions go to completion with minimal temperature overstep (0.13°C for a 10°C/m.y. event) provided a fluid transporting medium is present. Analysis of mineral–fluid equilibration rates during H_2O infiltration of carbonate rocks suggests that the latter can rapidly buffer the coexisting fluid composition. It appears, therefore, that calculated fluid–rock ratios for metamorphism of carbonate rocks are likely to correspond to those that actually occurred (Ferry; Wood and Walther, this volume).

Acknowledgments

We benefited from helpful comments by R. Wollast. We also thank Cheril Cheverton for typing the manuscript and drafting the figures. Financial support was provided in part by National Science Foundation grants EAR83-18905 and EAR82-12502.

References

Aagaard, P., and Helgeson, H.C. (1982) Thermodynamic and kinetic constraints on reaction rates among minerals and aqueous solutions. I. Theoretical considerations. *Amer. J. Sci.* **282**, 237–285.

Berner, R.A. (1978) Rate control of mineral dissolution under earth surface conditions. *Amer. J. Sci.* **278**, 1235–1252.

Busenburg, E., and Clemency, C.V. (1976) The dissolution kinetics of feldspars at 25°C and 1 atm CO_2 partial pressure. *Geochim. Cosmochim. Acta* **40**, 41–49.

Chou, L., and Wollast, R. (1984) Study of the weathering of albite at room temperature and pressure with a fluidized bed reactor. *Geochim. Cosmochim. Acta* **48**, 2205–2218.

Chou, L., and Wollast, R. (1985) Steady-state kinetics and dissolution mechanisms of albite. *Amer. J. Sci.* **285**, 963–993.

Eyring, H. (1935) The activated complex in chemical reactions. *J. Chem. Phys.* **3**, 107–115.

Ferry, J.M. (1984) A biotite isograd in South-Central Maine, U.S.A.: Mineral reactions, fluid transfer, and heat transfer. *J. Petrol.* **25**, 871–893.

Fleer, V.N., and White, W.B. (In press) The dissolution kinetics of alkaline earth feldspars in aqueous solutions at temperatures below 100°C. *Geochim. Cosmochim. Acta.*

Garrels, R.M., and Howard, P. (1959) Reactions of feldspar and mica with water at low temperature and pressure, in *Proc. Sixth Nat. Conf. on Clays and Clay Minerals,* pp. 68–88. Pergamon Press, Oxford.

Goldich, S.S. (1938) A study of rock weathering. *J. Geol.* **46,** 17–58.

Graham, C.M., Greig, K.M., Shepherd, S.M.F., and Turi, B. (1983) Genesis and mobility of the H_2O-CO_2 fluid phase during regional greenschist and epidote amphibolite facies metamorphism: A petrological and stable isotope study in the Scottish Dalradian. *J. Geol. Soc. London* **140,** 577–599.

Grandstaff, D.E. (1980) The dissolution rate of forsteritic olivine from Hawaiian beach sand, in *Proc. 3rd Internat. Symp. on Water–Rock Interaction,* pp. 72–74. Edmonton, Canada.

Helgeson, H.C., Murphy, W.M., and Aagaard, P. (1984) Thermodynamic and kinetic constraints on reaction rates among minerals and aqueous solution. II. Rate constants, effective surface area, and the hydrolysis of feldspar. *Geochim. Cosmochim. Acta* **48,** 2405–2432.

Holdren, G.R., Jr., and Berner, R.A. (1979) Mechanism of feldspar weathering. I. Experimental studies. *Geochim. Cosmochim. Acta* **43,** 1161–1171.

Jones, K.A., and Galwey, A.K. (1966) Size distribution, composition, and growth kinetics of garnet crystals in some metamorphic rocks from the west of Ireland. *Quart. J. Geol. Soc. London* **122,** 29–44.

Katchalsky, A., and Curran, P.F. (1965) *Nonequilibrium Thermodynamics in Biophysics.* Harvard University Press, Cambridge. 248 pp.

Krauskopf, K.B. (1967) *Introduction to Geochemistry.* McGraw-Hill, New York. 721 pp.

Kretz, R. (1966) Grain-size distribution for certain metamorphic minerals in relation to nucleation and growth. *J. Geol.* **74,** 147–173.

Kretz, R. (1973) Kinetics of the crystallization of garnet at two localities near Yellowknife. *Can. Mineral.* **12,** 1–20.

Lagache, M. (1965) Contribution a l'etude de l'alteration des feldspaths, dans l'eau, entre 100 et 200°C sour diverse pressions de CO_2, et application a la synthese des mineraux argileux. *Soc. Fr. Min. Crys. Bull.* **88,** 223–253.

Lasaga, A.C. (1981) Transition state theory, in *Kinetics of Geochemical Processes,* edited by A.C. Lasaga and R.J. Kirkpatrick, pp. 135–169. *Reviews in Mineralogy* 8. Min. Soc. Amer., Washington, D.C.

Lasaga, A.C. (1984) Chemical kinetics of water-rock interactions. *J. Geophys. Res.* **89,** 4009–4025.

Lin, F.-C., and Clemency, C.V. (1981a) The kinetics of dissolution of muscovites at 25°C and 1 atm CO_2 partial pressure. *Geochim. Cosmochim. Acta* **45,** 571–576.

Lin, F.C., and Clemency, C.V. (1981b) Dissolution kinetics of phlogopite. I. Closed systems. *Clays Clay Min.* **29,** 101–106.

Mazor, J.J., and Walther, J.V. (1985) Steady-state rate constants for silica glass as a function of pH at 65°C. Geol. Soc. Amer., *Abstracts with Programs* **17,** no. 7, 656.

Nash, V.E., and Marshall, C.E. (1956) The surface reactions of silicate minerals. Pt. I. The reactions of feldspar surfaces with acidic solutions. *Univ. Missouri Agr. Expt. Sta. Res. Bull.* **613,** 36 pp.

Rimstidt, J.D., and Barnes, H.L. (1980) The kinetics of silica-water reactions. *Geochim. Cosmochim. Acta* **44,** 1683–1700.

Rumble, D., III, Ferry, J.M., Hoering, T.C., and Boucot, A.J. (1982) Fluid flow during metamorphism at the Beaver Brook fossil locality. *Amer. J. Sci.* **282,** 866–919.

Schott, J., Berner, R.A., and Sjoberg, E.L. (1981) Mechanism of pyroxene and amphibole weathering. I. Experimental studies of iron-free minerals. *Geochim. Cosmochim. Acta* **45,** 2123–2135.

Spry, A. (1969) *Metamorphic Textures.* Pergamon Press, Oxford. 350 pp.

Tamm, O. (1930) Expermentelle Studien uber die Verwitterung und Tonbildung von Feldspaten. *Chemie de Erde* **4,** 420–430.

Tole, M.P., Lasaga, A.C., Pantano, C., and White, W.B. (In press) Factors controlling the kinetics of nepheline dissolution. *Geochim. Cosmochim. Acta* **50.**

Walther, J.V., and Orville, P.M. (1982) Rates of metamorphism and volatile production and transport in regional metamorphism. *Contrib. Mineral. Petrol.* **79,** 252–257.

Walther, J.V., and Wood, B.J. (1984) Rate and mechanism in prograde metamorphism. *Contrib. Mineral. Petrol.* **88,** 246–259.

Winchell, P. (1969) The compensation law for diffusion in silicates. *High Temp. Sci.* **1,** 200–215.

Wood, B.J., and Walther, J.V. (1983) Rates of hydrothermal reactions. *Science* **222,** 413–415.

Wynne-Jones, W.F., and Eyring, H. (1935) The absolute rate of reactions in condensed phases. *J. Chem. Phys.* **3,** 492–502.

Index